Observing Projects Using *Starry Night Backyard*™ and *Deep Space Explorer*™

Marcel W. Bergman

William J. F. Wilson

T. Alan Clark

W. H. Freeman and Company
New York

ISBN: 0-7167-7338-4
EAN: 97807167-7338-2

Printed in the United States of America

First printing

W. H. Freeman and Company
41 Madison Avenue
New York, NY 10010
Houndmills, Basingstoke RG21 6XS England

www.whfreeman.com

CONTENTS

PREFACE

Astronomy is an observational science. The opportunity to make meaningful observations of the real sky is vital to this study. However, for various reasons, such as light pollution or the lack of adequate equipment or resources, many students of astronomy are denied this opportunity. The projects described in this book provide an effective substitute for this observing experience (and also enhance the experience of students blessed with the resources to make real ground-based observations) by allowing students to make significant and accurate observations of the virtual sky using the *Starry Night Backyard*™ and *Deep Space Explorer*™ programs.

These programs act as virtual telescopes by allowing students to make observations that, although not perhaps possible in reality, nevertheless accurately convey current astronomical knowledge. Observations that require months of real time to accumulate can be made with a condensed time animation. Observational hindrances, such as daylight and the horizon, can be removed to make observations more easily understandable. These programs allow phenomena and events, which can be difficult to understand from the reference frame of the rotating and orbiting Earth, to be viewed from a simpler and more elegant frame of reference. This change of perspective often makes obvious an otherwise subtle or difficult concept.

The projects in this book allow students to examine the appearance and behavior of various astronomical objects interactively. Physical laws and astronomical concepts become more interesting when students demonstrate them through their own "observations." As examples, the reenactment of Romer's speed of light observations outlines the difficulty of this pioneering experiment, the observation of the 76-year orbit of Halley's comet underlines the crux of Kepler's Second Law, while the measurement of parallax makes cosmic distances less daunting and more wondrous. The projects guide students in their observations and their scientific analysis. Interspersed questions help students focus their observations upon particular concepts or theories. The beauty and wonder of the cosmos unfolds as students make observations of the planets and their satellites, the nearby stars, deep-sky objects, and finally the three-dimensional soap-bubble structure in the realm of galaxies, galaxy clusters and superclusters.

These observing projects present basic ideas at several levels. Most begin by introducing a basic idea in a simple way, using different viewing locations to illustrate this idea in the most effective manner. Then the simulations are extended to include measurements and simple analysis to demonstrate specific physical principles. Many of the projects use measurements that could not be carried out in real life. Some examine the universe from locations that are inaccessible to us but which provide insight into the underlying science. Several chapters in the book contain mathematical derivations, but these derivations always culminate in simple formulae to which the observations and measurements can be applied. Thus, the point of the observations can be demonstrated without the necessity of a full understanding of the mathematical derivations. Several exercises require graphing of the measurements in order to demonstrate a particular concept. This approach—the application of mathematical formulae and graphing to measured results—is designed to introduce students to the methodology of science.

The setting up of initial conditions to illustrate the various concepts of astronomy in *Starry Night Backyard*™ can often be quite complex and time consuming. Thus, configuration files for these initial conditions have been prepared for all projects. These files have been placed on a Web site and can be downloaded at the time of installation of the software. These files are readily accessed in *Starry Night Backyard*™ through the Go menu on the control panel. The availability of these files avoids the error-prone and time-consuming setup steps and allows students to focus more clearly upon astronomy rather than on software interface techniques.

While some of the projects in this book are relatively simple in concept and execution, others will appeal to more advanced students. Nevertheless, in all exercises, we have taken care to test and retest the setup files and the instruction sequences so that they are very complete. They should permit all students to carry out the tasks efficiently and reproducibly.

The Introduction outlines the purpose of the book and suggests preferences that optimize the program interface for the execution of these specific projects and ends with a simple project that emulates Clyde Tombaugh's discovery of Pluto in 1930.

A Glossary of Procedures augments the very complete User's Manuals on the *Starry Night Backyard*™ and *Deep Space Explorer*™ CDs and introduces specific features of these programs that are used throughout these projects. Students should read these manuals and the glossary as a prelude to carrying out the projects, and experiment with the many features of these programs before attempting the individual projects.

It is a pleasure to thank the staff of W. H. Freeman and Company for the opportunity to develop this book and for their careful and professional approach to it. We thank particularly Victoria Anderson for her guidance, constant encouragement, and professionalism throughout this project. We are very grateful for the work of Jodi Isman, as she and her team shepherded the book through the editing and production stages with skill and a constant concern for quality. We also thank Pedro Braganca and David Whipps at Imaginova for their help in streamlining the inclusion of the preconfigured setup files in this project. Colleagues at the University of Calgary, particularly David Fry and Eugene Milone, have been a constant source of ideas and suggestions. The authors are indebted to Jennifer Lowndes for the suggestion that the alignment of Brunel's Box Tunnel be included in an exercise investigating sunrise positions. Above all, the authors express their gratitude and thanks to their wives, Marcia, Dawn, and Jean, respectively, for their support during the development of this book.

INTRODUCTION

The observing projects in this book were designed and tested using *Starry Night Backyard*™ v.4.0.5 and *Deep Space Explorer*™. Prior to attempting any of the projects in this book, it is important that you become familiar with the interfaces of these programs. Review the User's Guides for both of these software applications. Spend some time experimenting with their interfaces.

A. Assumptions

We assume that you are familiar with the following techniques as they apply to your operating system:

- Standard mouse techniques;
- Menu navigation;
- File management including opening, saving, and copying files.

We also assume that you are familiar with the following techniques as they apply to *Starry Night Backyard*™ and *Deep Space Explorer*™:

- How to navigate the menu structure.
- How to open views from the Go/Observing Projects folder.
- How to set global preferences.
- How to use control panel features to change the date, time, viewing location, elevation, and field of view.
- How to open contextual menus for the sky or a particular object.
- How to center an object in the view.
- How to open, close and select options in the auxiliary panes.
- How to use of the Hand Tool and/or scrollbars for adjusting the direction of view.
- How to use the angular measurement feature of the Hand Tool to measure angular distances. Many of the projects require such measurements and it is a good idea to practice this technique.

Although we assume that you are familiar with these techniques, the Glossary of Procedures at the back of this book gives a brief outline of how to use the program features that you will need most often as you work through the Observing Projects.

B. How to Use This Book

In each Observing Project, you will make observations of the virtual sky in *Starry Night Backyard*™ and/or *Deep Space Explorer*™ that demonstrate a particular astronomical concept. While each project is self-contained, allowing completion of the projects in any order, the chapters in this book are organized to present progressively more difficult astronomical ideas.

Observations that require *Starry Night Backyard*™ generally use one or more files that set up the required initial parameters for a particular section of a project. These preconfigured setup files will need to be downloaded from the Web site www.whfreeman.com/universe 7e. Navigate the Web site to the Observing Projects page and click the button labeled **Download Setup Files**. This will install these files into the appropriate folder of your hard drive and make these files accessible from the Observing Projects item under the Go menu.

It is important that you use the Go command in the menu to select the preconfigured setup files, rather than opening them from the File menu of the *Starry Night Backyard™* interface. The latter route results in multiple files being opened, which slows down the operating speed of the program. (Unfortunately, when using the Go command, the names of the setup files are not displayed as they are used.)

It is also important that, when exiting the *Starry Night Backyard™* application, you click the "Don't Save" button in the prompt that appears. If you inadvertently save changes to any configuration, you can restore the original setup file back into the appropriate folder on your hard disk by reinstalling the files from the Web site. To avoid the problem of overwriting these setup files, you or your system administrator can change the attributes of the files in the Observing Projects folder to Read Only. This folder will be found in the subfolder Sky Data/Go in the folder containing your *Starry Night Backyard™* application.

After a brief introduction outlining the concept explored in the project, each chapter consists of sections that contain specific instruction sequences for making observations and measurements of the virtual sky. Instruction steps are numbered and shaded. Interspersed among instructions are descriptive paragraphs focusing on what the simulation demonstrates, and questions that help you to focus on the significance of your observations.

Although we do not give explicit instructions of how to invoke particular features of *Starry Night Backyard™* and *Deep Space Explorer™* from the program interface, when we have found a particularly efficient technique for accomplishing a function, we include it as a tip in an information box. You will also see references in an information box directing you to sections of the textbook, Freedman and Kaufmann, *Universe*, 7th Ed., that are relevant to the concepts described in the project. Finally, boxes within the text include occasional parenthetical remarks of interest, for example, describing how the observations being made in the simulation are accomplished in real life.

C. Sample Project: Discovering Planet X

In 1905, Percival Lowell, based on William Herschel's discovery of Uranus in 1781 and Johann Gottfried Galle's discovery of Neptune in 1846, speculated that still more planets might exist in the solar system beyond Neptune. Although Lowell died in 1916, the staff at the observatory that he founded in Flagstaff, Arizona continued the search. In 1930, Clyde Tombaugh photographed the sky in the direction of the constellation Gemini on two different nights.

1. Launch *Starry Night Backyard™*.
2. Select **File/Preferences.** Open the **Cursor Tracking (HUD)** page in the Preferences window and select only Name and Object type in the Show list.
3. Open the **Brightness/Contrast** page in the Preferences dialog window and slide the **Star Brightness** control all the way to the right.
4. Close the Preferences window.

TIP: You can use the keyboard shortcut Ctrl+Shft+P (Cmd+Shift+P on the Macintosh) to open the Preferences dialog window.

To simulate Tombaugh's observations, you will need to change the viewing location and time. In most of the projects, these steps will have been done for you in the setup file.

5. Change the viewing location to **Flagstaff, Arizona.**
6. Stop time flow.
7. Adjust the date to **January 23, 1930.**

8. Change the time in the control panel to 2:00:00 A.M. standard time.
9. Find the star **HIP35807**. The view is now centered upon this star.
10. Zoom in to a field of view between 40' and 50' wide.
11. Change the time step interval in the control panel to 6 sidereal days. Using an interval of sidereal days will keep the stars in the view stationary.

> **TIP:** See the Glossary of Procedures for details on how to change the viewing location, adjust the date and time, and find objects in the sky.

The view shows a part of the field of stars that Clyde Tombaugh photographed on January 23, 1930. Tombaugh photographed the same region of the sky 6 days later on January 29, 1930.

12. Observe this view, carefully noting the positions of all of the objects in the view.
13. Click the single step forward button once.

> The **sidereal day** is discussed in Box 5-2 of Freedman and Kaufmann, *Universe*, 7th Ed.

While the stars remain stationary when time advances 6 sidereal days, you might have noticed that one object in the view has moved in this time interval. Clyde Tombaugh used an instrument called a blink comparator to discover the movement of this object against the background stars. The blink comparator allows the observer to switch rapidly between two precisely aligned photographic plates of the same region of sky taken at different times. Observing the images through a microscope, the operator of the device can spot objects in the field that have moved in the time interval over which the two photographs were taken. The next steps allow you to simulate the operation of a blink comparator.

14. Alternately step **Time Backward** and **Forward** by one step of 6 sidereal days to simulate a blink comparator.
15. Use the **Hand Tool** to identify the object that moves between views.

Question 1: Which object did Clyde Tombaugh discover in 1930?

16. Open the contextual menu over the object that Tombaugh discovered and choose Center from the menu.
17. Zoom in to a field of view of about 3" to see a magnified image of this object.
18. Identify the other object in the view.

Question 2: Which other object appears in the view?

19. Open the contextual menu and select **Show Info** over both of the objects in the view to learn more about them.

Question 3: What is the angular size of the object that Tombaugh found? (*Hint:* Look under the Other Data layer of the Info pane.)

D. Preferences

During the development of these projects, we have found that some of the options available in *Starry Night Backyard*™, while interesting, can interfere with quantitative observations or reduce the efficiency of observations. The following steps adjust these preferences.

20. Select **File/Preferences**.
21. Open the **Responsiveness** page of the Preferences dialog window.
22. Deselect the options **Pan to found objects** and **Animate changes in location**.
23. Open the **Brightness/Contrast** page and reset the Star Brightness control to the default level.
24. Close the Preferences dialog window.
25. Open the **Find** pane.
26. Click the down arrow next to the Sun, move to the **Halo Effects** item in the menu, and choose **Never**. Also, uncheck the **Lens Flare** option.
27. Select **View/Save View Options as Default** from the menu.
28. Exit *Starry Night Backyard*™ without saving changes to the file Planet X.

E. Conclusion

This introduction has informed you of the general format of the projects in this book. It has allowed you to familiarize yourself somewhat with the *Starry Night Backyard*™ interface by having you reproduce the historic observations of Clyde Tombaugh. Finally, it has configured the preferences and options of the *Starry Night Backyard*™ software to make it most efficient and reliable for the projects in the remainder of this book.

STARS AND CONSTELLATIONS 1

This project will familiarize you with several of the major constellations and bright stars that can be seen from midnorthern latitudes on the Earth. When you know these patterns and can find them in the nighttime sky, you can use them to locate other stars and constellations. With practice, you can become expert at finding your way around the sky.

Historically, a constellation is a group of stars that outlines, or at least represents in some fashion, a familiar object or pattern. For example, the bright stars in the constellation of Orion roughly outline the body of Orion, the hunter, while the long line of stars making up the constellation Draco resemble the curved shape of a dragon. Many of these names have their origins in the myths and stories of antiquity.

In modern astronomy, constellations are defined differently. Each constellation is an area of the sky with precisely defined boundaries, rather than a group of stars. Together, the 88 officially designated constellations cover the entire celestial sphere. In this way, the constellations provide us with an atlas of the celestial sphere. Thus, when we say that the star Regulus is in the constellation of Leo, this tells us the location of Regulus on the sky in the same way that saying that Grenoble is in France tells us where Grenoble is upon the Earth. However, most people—including most astronomers—picture a constellation more often in terms of the historical star pattern than as a set of invisible boundary lines dividing the sky into sections.

> Section 2-2 of Freedman and Kaufmann, *Universe*, 7th Ed., discusses the patterns of stars that form the constellations.

A. Constellations

In this section, you will use the *Starry Night™* program to understand the difference between the historical and modern definitions of constellations.

1. Launch *Starry Night™*.
2. Select **File/Preferences** from the menu. In the Preferences dialog choose **Cursor Tracking (HUD)** in the drop box. In the **Show** list, choose only the options **Constellation common name, Name,** and **Object type**.
3. Stop time flow and adjust the time to 12:00:00 A.M. Standard Time, at your home location.
4. In the **View Options** pane, expand the **Constellations** layer and click over the item named **Stick Figures** to open the Constellation Options dialog window. In the Constellation Options dialog window check the Stick Figures and choose Astronomical in the Kind drop box. Check the Labels options and then click OK.

The view shows the familiar historical definition of the constellations. Stars within a constellation are grouped to form a recognizable pattern. Scroll across the sky to survey the various constellations visible from your home location.

> 5. In the **View Options** pane, turn off the **Stick Figures** option under the **Constellations** layer and check the **Labels** and **Boundaries** options.

Now the sky is divided by boundaries like countries on a map. Each "country" shows the modern astronomical definition of a constellation. A constellation contains all of the stars within its boundaries regardless of whether these stars are also included in the more familiar historical pattern of the constellation.

The brighter stars within a constellation are generally ranked according to their apparent brightness with Greek letter prefixes. For example, the brightest star in the constellation Leo, the Lion, is called *alpha Leonis;* the next brightest star is called *beta Leonis;* and so on. Often, these prominent stars will also have common names. *Alpha Leonis,* for instance, is also known as Regulus.

A third display option is available, showing an artist's impression of imaginary images representing the constellations.

> 6. In the **View Options** pane under the **Constellations** layer, toggle on the **Illustrations** option and click off the **Boundaries** option.

You can again scroll across your local sky to view these images. These striking representations may help you in remembering the form of the constellations when you come to explore the real sky with none of these guides available.

The table below shows the optimum times for looking for major constellations or well-known collections of stars called **asterisms** from midlatitude northern hemisphere sites.

Constellation or Asterism	Time of Best Visibility
Big Dipper	February to August
Cassiopeia	August to March
Auriga	November to April
Bootes	April to August
Virgo	April to July
Leo	March to June
Orion's Belt	December to April

B. Asterisms

Some prominent patterns of stars that have common historical names are not constellations. These are called **asterisms**. Some asterisms are part of a constellation. An example is the Big Dipper asterism, which forms a part of the constellation Ursa Major. Other asterisms extend over two or more constellations. For example, the Summer Triangle asterism consists of three stars, each from a different constellation. The next steps will help you to understand the difference between asterisms and constellations.

7. Select the view named **Constellations B** under **Go/Observing Projects/Stars and Constellations**.

The view shows the night sky as seen looking west about 45° above the horizon from Calgary, Canada, at midnight on June 4, 2006. Constellation patterns, labels and modern boundaries are visible. Notice Ursa Major, commonly known as the Big Bear, to the right of center in the view.

8. Click the **Stick Figures** item under the **Constellations** layer of the **View Options** pane to open the **Constellation Options** dialog window. In the dialog window, check the **Stick Figures** option choosing **Asterisms** as the **Kind** of figure to illustrate. Uncheck the **Boundaries** option and check the **Labels** option. Then click the **OK** button.

Now the view shows the asterisms that are found in this part of the sky. You will see that the Big Dipper asterism, so named because its shape resembles that of a scoop or dipper, is within the boundaries of Ursa Major but forms only a part of the constellation's more extensive pattern. Notice also the large asterism named the Diamond of Virgo. Each of the four stars in this asterism comes from a different constellation.

9. Toggle back and forth between the options **Asterisms** and **Astronomical** in the Constellation Options dialog window to answer the following question.

Question 1:   Which constellations contribute stars to the Diamond of Virgo asterism?

As the Earth rotates, the orientation of constellations and asterisms varies with respect to compass directions upon Earth. Their visibility will depend upon your location upon the Earth and the season. Nevertheless, their easily recognizable patterns make asterisms and constellations invaluable guides for finding your way in the night sky.

C. Finding the Big Dipper

Figure 2–6 in Freedman and Kaufmann, *Universe*, 7th Ed., shows the Big Dipper and demonstrates its value as a guide to other stars and constellations.

A very useful guide to the sky for Northern Hemisphere observers is the above-mentioned Big Dipper asterism within the constellation Ursa Major, the Great Bear. *An asterism is a pattern of a few stars within a constellation or extending across several constellations that help in identifying regions of your sky.* This asterism is in the shape of a spoon or scoop that might be used to serve soup or scoop water from a barrel. It is made up of seven bright stars. Three of them in the tail of the bear make up the handle while the other four make up the scoop and are in the body of the bear. If you live between latitudes of about 40° N and 60° N and want to find the Big Dipper asterism in the actual sky, face approximately north and look at a point about halfway up from the horizon. Imagine a large circle centered on this point, starting near the northern horizon and curving up to the left to a point near the zenith, and then curving down to the right, back to the starting point. The easily recognized pattern of the Big Dipper should be somewhere along or near this circle. If you live south of 40° N latitude, the Big Dipper may be below the horizon during certain times of the year or at certain times of the night. Consult the previous table for the best times to view the Big Dipper. If you live north of 60° N latitude, the imaginary circle described above will include points south of the zenith.

The Big Dipper can have any orientation, including upside down or hanging downward from its handle, depending on the time of night and the time of year. The important thing is to be able to recognize its pattern.

10. Select the view named **Constellations C** under **Go/Observing Projects/Stars and Constellations.**

The view is of the northern sky from New York City at 10:00 P.M. EDT on September 1, 2006. Using the technique described above, you should quickly find the Big Dipper near the left side of the view. The seven bright stars that form the shape of a dipper comprise the Big Dipper. At this time, the handle extends away from the bowl towards the upper left of the screen.

11. To confirm that you have correctly identified the Big Dipper in the view, open the **Constellation Options** dialog window and select **Stick Figures, Asterisms, and Labels.** When you have identified the Big Dipper in the view, turn the **Labels** option off again under the **Constellations** layer of the **View Options** pane.

D. Describing Directions from the Big Dipper

The Big Dipper is an important asterism to locate because its member stars can be used as pointers to many other stars and constellations. For example, two of this asterism's stars provide a pointer to the North Star. The ability to find the North Star is very useful because it always lies directly above the north point of the horizon, at least at this time in history—a valuable guide when orienting oneself on the Earth.

In order to use the Big Dipper as a guide to other constellations, asterisms, and stars, we first need to define the directions "upward" and "downward" relative to the Big Dipper. Imagine that the dipper shape is oriented exactly horizontal with the bottom of the bowl resting on a flat table and the handle extending toward your left. In this orientation, the bowl opens upward and it is in this sense that we shall use the term "upward." Conversely, downward is the direction in which water would drip if the bowl of the dipper leaked. Remember, the bowl of the Big Dipper always opens "upward" toward the North Celestial Pole no matter which orientation the Big Dipper has in the sky.

12. Adjust the time in the control panel to **12:00:00 A.M.** in order to see the orientation of the Big Dipper as described in the previous paragraph.
13. Turn the **Stick Figures** option off under the **Constellations** layer of the **View Options** pane.
14. Start **Time Flow Forward** and watch the changing orientation of the Big Dipper in the night sky as time progresses at 3000× its normal rate. **Stop Time Flow** at approximately **5:00 A.M.** on September 2.

With time flowing forward at this advanced rate, you see the change in the orientation of the Big Dipper with respect to the horizon as the sky rotates around the North Star due to Earth's rotation. By 5:00 A.M., the Big Dipper is standing on its handle on the right side of the view. In this orientation, "upward" from the bowl, as defined above, is toward the upper left in the sky and "downward" is toward the lower right.

It may seem odd to use the word "upward" in this sense; however, it allows us to use the Big Dipper as a guide to the sky without worrying about how the Big Dipper is oriented with respect to the horizon.

E. Finding the North Star

To find Polaris, the North Star, using the Big Dipper as a guide, first locate the handle of the Big Dipper and then find the two stars that are on the end of the bowl farthest from the handle. These two stars are often called the "pointer stars" because they point to the North Star.

15. Select **Go/Observing Projects/Stars and Constellations/Constellation C** to return to the original time.
16. Position the **Hand Tool** over the bottom pointer star (the star at the bottom of the bowl farthest from the handle). Use the angular measurement facility of the **Hand Tool** to draw a line directly through the top pointer star (the star at the top of the bowl) and extend the line a distance approximately equal to seven times the distance between the two pointer stars, or about 35°.
17. Release the mouse button to deactivate the angular measurement feature and position the **Hand Tool** over the nearest bright star to display its name, Polaris, the North Star.

The pointing is not perfect but it is fairly close. While the North Star is not a really bright star, it is the brightest star in that little patch of the sky. At this time in history, this star is close to the North Celestial Pole, the point in the sky directly above the north end of the spin axis of Earth.

F. Finding the Little Dipper

Polaris is the end star in the handle of the Little Dipper, an asterism in the constellation Ursa Minor. Starting from Polaris, you should be able to discern a line of four stars curving in an arc toward the handle of the Big Dipper. The first three of these stars are fainter than Polaris and the fourth is about equal in brightness to Polaris.

18. Select **Stick Figures** and **Labels** under the **Constellations** layer in the **View Options** pane to verify that you have found the **Little Dipper**. Then turn off the **Stick Figures** and **Labels** options.

At the end of the Little Dipper nearest to the handle of the Big Dipper are two stars, Kochab and Pherkad. These two stars, which are brighter than all of the other stars in the Little Dipper except for Polaris, are often referred to as the Guardians. As the sky turns, the Guardians always remain between the Big Dipper and Polaris and "guard" Polaris from the Great Bear.

19. Use the HUD feature of the **Hand Tool** to identify the two Guardian stars, Kochab and Pherkad.

G. Finding Cassiopeia, or the "W"

20. Select **Go/Observing Projects/Stars and Constellations/Constellations C** to return to the initial settings.

At this time, with the Big Dipper to the west of north in the view, the constellation Cassiopeia is found by imagining a straight line starting at the handle of the Big Dipper to the North Star and then extending the line approximately an equal distance past the North Star. The brightest stars in the constellation Cassiopeia (the Queen of Ethiopia) form an asterism shaped like a W. The end of this line should place you within this group of stars. In the orientation on your screen, the W is tipped up on one end.

21. Turn on the **Stick Figures** and **Labels** options under the **Constellations** layer in the **View Options** pane to verify that you have found the W asterism. Then turn off these options.

H. Finding the Star Capella and the Constellation Auriga

Capella is the brightest star in the constellation Auriga, the Charioteer. To find this star, first locate the two stars on the top of the bowl of the Big Dipper. One of these two stars, Megrez, marks the point where the handle joins the bowl, and the other is the upper of the two pointer stars (Dubhe).

22. Beginning at the star Megrez, use the angular measurement feature of the **Hand Tool** to draw a line through Dubhe and extend it across the sky (toward the right on your screen) a distance of about six times the spacing between Megrez and Dubhe (about 60°). The REALLY bright star near the end of this line is Capella. Use the HUD feature of the **Hand Tool** to verify your identification of this star.

Capella is actually slightly above the imaginary line you have drawn in the previous step, but it is the only really bright star that stands out in that part of the sky. When you look at Capella in the actual sky you may notice that it has a yellowish color. This is because Capella is roughly the same temperature as the Sun. However, while our Sun is a dwarf star, Capella is a supergiant star with a radius approximately 14 times larger than that of the Sun and is therefore intrinsically much brighter than the Sun.

The brightest stars of the constellation Auriga consist of five stars, including Capella, in the form of a stretched pentagon.

23. Center the view on Capella and allow time to run forward to about 1:00:00 A.M. on September 2, 2006.
24. Open the Constellation Options dialog window and select **Stick Figures, Astronomical,** and **Labels.** Next, click the **OK** button in order to see the pattern of the constellation Auriga. You can also turn on the **Illustrations** briefly to see an artist's impression of this constellation.

I. Finding the Star Arcturus and the Constellation Bootes

25. Select **Go/Observing Projects/Stars and Constellations/Constellations C** menu to return to the initial configuration.
26. Scroll the view so that the gaze direction is between W and NW with the horizon low on the screen.

In this view, the Big Dipper is visible in the upper right quadrant of the screen. To find the star Arcturus, start by looking at the handle of the Big Dipper. Notice that the handle is curved and forms an arc in the sky. Imagine continuing this arc through the sky in a direction away from the bowl of the Big Dipper for a distance roughly equal to the total length of the Big Dipper asterism. You should discover that this arc passes more-or-less through a bright star above and to the right of the center of the view. This star is Arcturus. A good way to remember how to find this star is, "Follow the arc to Arcturus."

Arcturus is the brightest star in the constellation Bootes, the hunter. (The two "o's" in Bootes are pronounced separately: Bo-otes, which is written as Boötes. The "e" is also pronounced.) If you look up and right from Arcturus in the view you should see a pentagon of five stars. Arcturus plus the other five form a kite or ice cream cone shape in the sky.

> 27. Turn on the **Stick Figures** and **Labels** options under the **Constellation** layer of the **View Options** pane to check that you have found Bootes correctly.

In the real sky, Arcturus has a yellowish or even slightly orange tint because it is slightly cooler than the Sun. It is also a giant star—much smaller than the supergiant Capella, but still much larger than our Sun.

With the constellation lines and labels on in the view, you can see another constellation, Canes Venatici, below the handle of the Big Dipper. Canes Venatici (Latin for "dogs of the hunter") consists of two relatively faint stars joined by a single straight line. In mythology, Boötes is hunting the Great Bear. As the sky turns counterclockwise around the North Star, Boötes follows the Great Bear around the sky. The two stars in Canes Venatici are the two hunting dogs belonging to Boötes, nipping at the heels of the Great Bear as it circles the North Star, trying to get away from Boötes.

> 28. Turn on the **Illustrations** option under the **Constellations** layer of the **View Options** pane to see an illustration of these mythological figures. Turn this feature off again after you have examined these mythical ancient figures.

J. Finding the Star Spica and the Constellation Virgo

> 29. Turn off the **Labels** and **Stick Figures** options under the **Constellations** layer of the **View Options** pane.
> 30. Change the time in the control panel to 8:30:00 P.M. on **September 1, 2006.**
> 31. Adjust the view so that you are looking due west with the horizon near the bottom of the screen.

In this view of the twilight sky, locate the Big Dipper in the upper right quadrant of the screen. To find Spica, start at the handle of the Big Dipper and *follow the arc to Arcturus!* Continue this arc for roughly the same distance again past Arcturus to *"speed on to Spica."* Spica is the relatively bright star near the horizon between the south and southwest compass points.

> 32. Use the **HUD** feature of the **Hand Tool** to identify Spica.

Spica is the brightest star in the constellation Virgo, the Virgin. Spica is much hotter than our Sun and when viewed in the actual sky, it can be seen to have a slightly bluish tint.

> 33. Activate the **Labels** and **Stick Figures** options under the **Constellations** layer of the **View Options** pane to see the pattern of stars in the constellation Virgo. Then turn off these features.

K. Finding the Star Regulus and the Constellation Leo

In order to find Regulus in the constellation Leo, the Lion, we need to move to December and change the time and viewing direction.

> 34. Change the date in the control panel to **December 1, 2005** A.D.
> 35. Adjust the time to **1:00:00** A.M., Daylight Saving Time off.
> 36. Change the view to midway between E and NE with the horizon near the bottom of the screen.

In this view, locate the Big Dipper, standing on its handle on the left side of the screen. To find the star Regulus, locate the two stars in the Big Dipper on the opposite side of the bowl from the pointer stars (i.e., the star where the bowl of the dipper meets the handle, Megrez, and the star Phecda to the right of Megrez).

> 37. Starting at Megrez, use the angular measurement feature of the **Hand Tool** to draw a line in the direction of Phecda and extend it a distance about 10 times the distance between Megrez and Phecda (about 50°). The bright star at the right-hand end of a small grouping of stars is Regulus. Use the constellation display and HUD features to verify your identification of the constellation Leo and its most prominent star, Regulus.

Regulus, like Spica, is hotter than our Sun and has a bluish tint when viewed in the real sky. Regulus is the brightest star in the constellation Leo. The most easily recognized part of this constellation is a line of stars running toward the upper left from Regulus forming the shape of a sickle or backward question mark.

> 38. To see the constellation **Leo** containing the Sickle asterism, open the Constellation Options dialog window, select **Stick Figures, Asterisms,** and **Labels** and then click **OK**.

The Sickle asterism represents the head of the lion in the constellation Leo. The hind end of the lion is formed by a triangle of stars below and to the left of the Sickle.

> 39. To see the rest of the constellation Leo, open the Constellation Options dialog window and select the option **Astronomical** in the drop box labeled **Kind** and click **OK**.
> 40. Select the **Illustration** option under the **Constellations** layer of the **View Options** pane to see an illustration of the mythical lion Leo.

L. Conclusion and Suggested Extensions

This project has shown you how to find your way around the sky by using the Big Dipper as a guide. You should use the techniques discussed in this project to locate these stars and constellations in the real sky. Then, using a star atlas or printouts from your *Starry Night*™ software and maybe a pair of binoculars, explore other constellations to extend your knowledge of the night sky.

CHANGING LATITUDE

The ancient Greeks knew that as they traveled south they would see stars above the southern horizon that could not be seen from Greece. Conversely, when they traveled north, stars that they could see in Greece no longer rose above the southern horizon.

The Greeks realized that this is exactly what would happen if the Earth were round. On a flat Earth, the same stars should be above the horizon for all observers, regardless of where they were located. On a round Earth, on the other hand, our view of the sky changes when we move to different latitudes.

In this observing project, you can learn how our view of the stars and constellations changes as we move to different latitudes in the northern hemisphere.

A. The View to the South

1. Launch *Starry Night*™ and select the view **Mexico City** from **Go/Observing Projects/Changing Latitude**.
2. Select **File/Preferences** from the menu. Choose **Cursor Tracking (HUD)** in the upper left drop box and select **Altitude, Name,** and **Object Type** from the **Show** list. Then close the Preferences dialog window.

> **TIP:** Check the option to display HUD data in the upper left corner of the screen rather than next to the chosen object in the **Preferences** dialog if you prefer.

This view shows the sky looking south from Mexico City, Mexico (latitude 19° N), at 1:27:01 A.M. Standard Time on April 24, 2005.

The constellation **Scorpius** appears just above and to the left of center in the view. The top end of Scorpius in this view is shaped like a letter "T" tilted over on its right side, and the bottom end is a tail shaped like a fishhook. The bright red star is **Antares,** a cool but very bright supergiant star that can be used as an aid to the identification of Scorpius in the real sky. The labeled star at the center of the "T" is Dschubba, which we shall use as a reference star in the next series of observations.

3. Check your identification of **Scorpius** by selecting the **Labels** and **Stick Figures** options under the **View Options/Constellations** menu.
4. Select **Local Meridian** and **Show Compass Direction** from the **View Options/ Guides** menu.

Notice that the reference star, Dschubba, is due south in this view, transiting the meridian. Therefore, you are seeing it at its highest position in your sky. This occurs because the stars and constellations rise in the eastern part of the sky and set in the western part, just as the Sun does. They are highest in the sky when they are half way between rising and setting (i.e., when they are due south as seen from the northern hemisphere).

5. Position the **Hand Tool** over Dschubba and note its altitude above the horizon in the HUD. Record this value in Data Table 1 below.
6. Open the **Info** pane and expand the **Location** layer. Note the latitude of Mexico City in Data Table 1. Then close the **Info** pane.
7. Select the **View Options** pane, expand the **Constellations** menu, and deactivate the **Labels** and **Astronomical** options.
8. Select the **View options** pane, expand the **Guides menu,** and turn off the **Local Meridian.**

In order to see how our view of the sky changes with latitude, we will move progressively northward and note the altitude of Dschubba when it transits the meridian at each new location.

9. Select the view **Chicago** from **Go/Observing Projects/Changing Latitude.**
10. Position the **Hand Tool** over Dschubba and note its altitude relative to the horizon as seen from Chicago. Record this value, and the latitude of Chicago obtained from the **Info** pane in the **Location** menu, in Data Table 1 below.
11. To answer the following questions, you may want to switch back and forth between the views from Mexico City and Chicago. To do so, alternately select these views using **Edit/Undo "Go" Item** and **Edit/Redo "Go" Item** from the menu.

TIP: You can use the keyboard shortcut Ctrl+Z to Undo an action and Ctrl+Shift+Z to Redo an action.

Question 1:   In which direction does the constellation Scorpius move with respect to the south horizon as the viewing location moves north from Mexico City to Chicago?

Question 2:   Why is the time at which Dschubba transits the meridian different for the two locations?

12. Open the file **Regina** from **Go/Observing Projects/Changing Latitude.** Note the date and time for the new location.
13. Position the **Hand Tool** over Dschubba and note its altitude above the horizon at the time it transits the meridian at Regina, Canada. Record this value, and the latitude of Regina from **Info/Location**, in Data Table 1.

Question 3:   Where is Scorpius with respect to the horizon on the screen, as seen from Regina? Is any part of Scorpius missing?

14. Open the file **Fairbanks** from **Go/Observing Projects/Changing Latitude** and note the altitude of Dschubba and the latitude of Fairbanks in Data Table 1.

DATA TABLE 1

Location	Latitude	Altitude of Dschubba
Mexico City, Mexico		
Chicago, USA		
Regina, Canada		
Fairbanks, USA		

15. To help answer the following questions, select the views **Mexico, Chicago, Regina,** and **Fairbanks** in turn under **Go/Observing Projects/Changing Latitude.**

Question 4: Where is Scorpius on the view as seen from Fairbanks? What part of Scorpius is visible? Where is the rest of Scorpius, relative to the horizon?

Question 5: For which of the cities in Data Table 1 is Scorpius completely above the horizon? (We are referring to the "constellation" here in the historical sense, as the pattern of bright stars shaped like a T and fishhook, rather than as the entire astronomical area of sky.)

Question 6: At about what latitude does the tail of Scorpius begin to disappear from view (i.e., NEVER rises above the horizon) for an observer traveling north?

Question 7: Use the observed positions of Scorpius on the screen to estimate the approximate latitude above which NO part of Scorpius is ever visible.

Question 8: Which of the following statements correctly describes the relationship between latitude and the elevation angle between the reference star and the horizon?
a) As latitude increases, the reference star's elevation angle increases.
b) There is no relationship between the reference star's elevation angle and latitude.
c) As latitude increases, the reference star's elevation angle decreases.

Question 9: Suppose that a scientist in Fairbanks has applied for a research grant to study a globular cluster of stars near the middle of Scorpius, using an observatory just outside Fairbanks. If you were on the granting agency, what would be your response to this proposal? Would your response change if the research proposal included a request for funds to travel to an observatory in Texas?

B. Pole Star Altitude and Latitude on the Earth

16. Select the view **Minneapolis** under **Go/Observing Projects/Changing Latitude.**

> **TIP:** Use the **Labels** and **Astronomical** options in the **View Options/Constellations** if necessary. Turn these options off again when you have identified Cassiopeia and the Big Dipper.

The view is now northward from Minneapolis, Minnesota at 7:30 P.M. Standard Time on October 1, 2005. Polaris, the North Star is labeled. Time flow is frozen and the time step is set at 3000 times the rate of real time. Identify the constellations Ursa Major, which contains the Big Dipper, in the lower left quadrant of the view, and Cassiopeia in the upper right.

17. Click the **Flow Time Forward** button, and watch what happens in the sky.

Question 10: What happens to stars on the left side of the view, including the Big Dipper, as time progresses? E.g., do they move upward or downward?

Question 11: What happens to stars on the right side of the view, including Cassiopeia, as time progresses? E.g., do they move upward or downward?

Question 12: Is there one star that appears to remain at rest, while all other stars move in circles around it? Which star is this?

As the sky rotates around the pole, watch what happens to the Big Dipper, especially between about 10 P.M. and 1 A.M. In the lower left side of the view, stars are setting below the horizon, and on the lower right side of the view, stars are rising above the horizon. From Minneapolis however, the Big Dipper never sets. It approaches the northern horizon, but passes above this horizon without setting, and then gets higher in the sky again.

When the time display shows about 1:30 A.M., as the Big Dipper is starting to rise again, you should see a very bright star near the lower left corner of the view. This is the second brightest star in the northern hemisphere, **Vega**, in the constellation **Lyra**, the Lyre, a bright, blue main-sequence star.

Question 13: What happens to Vega at about 3:45 A.M.?

Stars or constellations that move in circles around the Pole without ever setting are called **circumpolar**. The Big Dipper is circumpolar as seen from Minneapolis, while Vega is not.

18. Select **Go/Observing Projects/Changing Latitude/Minneapolis** from the menu to return to the initial settings.
19. Position the cursor over Polaris and, from the HUD, note its altitude above the northern horizon. Record this value in Data Table 2 below.
20. From the **Location** menu under the **Info** pane, note the latitude of Minneapolis in Data Table 2.

Question 14: How does the elevation angle of the Pole Star, Polaris, compare to the latitude of Minneapolis? (These numbers should be close but will not be precisely equal because the Pole Star is actually about a degree away from the position of the North Celestial Pole in the sky.)

21. Select the view **Houston** under **Go/Observing Projects/Changing Latitude**.

The view is to the north from Houston, Texas at the same time and date as the previous view from Minneapolis.

22. Note the altitude of **Polaris** above the northern horizon from Houston and record this value in Data Table 2, below. From the **Location** page under the **Info** pane, find the latitude of Houston and enter this value in Data Table 2.

DATA TABLE 2

Location	Latitude	Pole Star Altitude
Minneapolis		
Houston		

Question 15: How does the elevation angle of the Pole Star from this location compare with the latitude of Houston?

23. Click the **Flow Time Forward** button, and watch what happens in the sky.

Notice that all of the stars of the Big Dipper except the one at the top right corner of the bowl (the one closest to the Pole Star) set below the horizon, then rise again later.

Question 16: Out of all the stars in the sky, how does the number that is circumpolar as seen from Houston compare with the number as seen from Minneapolis (e.g., are more stars circumpolar as seen from Houston, fewer, etc.)?

24. Select the view **Equator** under **Go/Observing Projects/Changing Latitude.**
25. Start **Time Flow Forward.**

Question 17: Looking north from the Equator, are any stars circumpolar?

Question 18: Is Polaris circumpolar as seen from the equator? Why or why not?

26. Select **North Pole** from **Go/Observing Projects/Changing Latitude.**
27. Start **Time Flow Forward.** As time flows, use the **Hand Tool** to scroll the view eastward through a full 360°.
28. Drag the view downward to bring Polaris into view near the zenith.

Question 19: From the North Pole, are there any stars that are **not** circumpolar?

Question 20: Based on your observations, which one of the following statements do you think is correct?
a) The angle of the Pole Star above the horizon equals your latitude.
b) The angle of the Pole Star above the horizon equals 90° minus your latitude.
c) The angle of the Pole Star above the horizon does not depend on your latitude.

Question 21: If you were south of the equator, say in Australia, would you expect to see circumpolar constellations anywhere in the sky? If so, in what part of the sky would they be?

C. Simple Celestial Navigation

Imagine that you have been selected by the National Geographic Society to lead an expedition recreating Columbus's historic 1492 voyage to the New World. Your only navigational aids are a compass and a quadrant, a device used to measure the elevation angle (altitude) of stars and planets.

Like Columbus, you leave Palos de la Frontera, Spain, on August 3 and use dead reckoning and coastal navigation to reach La Gomera, Spain, in the Canary Islands. There you take on supplies and effect any necessary repairs to your ships.

On September 6, you set sail from La Gomera (latitude 28° 6' N) to cross the Atlantic Ocean, intending to meet the sponsors of the expedition in Nassau, Bahamas (latitude 25° 3' N), sometime in October.

29. Close the **Info** pane.
30. Select the view **La Gomera** under **Go/Observing Projects/Changing Latitude**.
31. Start **Time Flow Forward** and observe the night sky, seeking stars and constellations that will aid you in your navigation.

You wish to maintain a heading slightly south of due west without straying too far north or south. Unfortunately, four days into the voyage, a five-day storm blows you off course, leaving you lost. Nine days after leaving La Gomera, the night sky is finally clear. You need to determine whether to correct your subsequent heading toward the north or south in order to reach Nassau.

32. Open the file named **Lost** in the **Go/Observing Projects/Changing Latitude** folder.
33. Start **Time Flow Forward** and observe the sky through the night.

Question 22:   To get back on course, do you need to adjust your heading to the south or north of west after surviving the storm?

D. Conclusions

In this project, you have seen that the view of the southern sky becomes more and more restricted as you move northward in the northern hemisphere until, near to the poles, only about one-half of the celestial sphere is visible. Related to this, you have seen that more and more of the stars on this celestial sphere are circumpolar, that is, they can be seen throughout the night from any observing site.

Of course, the same would be true in the southern hemisphere as one moved farther south toward the southern polar regions until, from the South Pole, only one-half of all stars would be visible. However, from this site, these would be in the opposite hemisphere from the stars visible from the North Pole.

The other major fact that you have verified is that the angle between the horizon and the North Celestial Pole, approximated at this time in history by the Pole Star, is equal to the latitude of the observing site. In the final section of this project, you had the opportunity to discover how early navigators made use of this fact.

DIURNAL MOTION

3

E very day, the Sun rises in the east and moves (or appears to move) toward the west across the sky, finally setting below the western horizon. The Moon and stars also rise in the east and set in the west. This apparent westward motion of the sky over the course of a day or night is referred to as **diurnal motion** and is caused by the Earth's eastward rotation on its axis. If you are unsure of this, a simple experiment may convince you that it is true. Try turning yourself slowly toward the left, keeping your head pointing forward, while you watch your surroundings. Every object that starts on your left will appear to move slowly toward your right as you turn; that is, objects appear to move in the opposite direction to that in which you are turning. The converse is also true. If, when you are turning, you see objects shifting toward your right, then you must be turning toward the left, opposite to the apparent direction of motion that you see. In similar fashion, the reason that we see the Sun rise in the east each day and move slowly westward until it sets is that the Earth is turning toward the east.

> Section 2–3 of Freedman and Kaufmann, *Universe*, 7th Ed., discusses diurnal motion.

The Sun is only one of many objects in the sky. At night with the unaided eye, we can see the Moon and as many as five planets (Mercury, Venus, Mars, Jupiter, and Saturn) along with a lot of stars. Many other objects can be seen through binoculars or a telescope. Because it is the Earth and not the rest of the universe that is rotating, we should see all of these objects taking part in diurnal motion. However, with the exception of the Moon, we normally see these objects only at night. Because most of us spend very little time outside at night (until we become interested in astronomy!), we may be less likely to notice that stars rise and set. In fact, they do, just like the Sun.

In this project, you will observe and measure the diurnal motion of the Sun and other celestial objects.

A. The Sun's Diurnal Motion—Midday

1. Launch *Starry Night*™ and select the file **Diurnal A** from **Go/Observing Projects/Diurnal Motion**. Take a moment to review the settings.
2. Select **File/Preferences** and set **Cursor Tracking (HUD)** options to include **Name** and **Object type**.

In this initial view toward the south from Seattle at noon on March 1, 2005, the Sun is almost due south. To the right (west) and slightly below the Sun you should be able to see a starlike object. This is the planet Venus.

You may have expected that the Sun would be directly south at noon, and therefore located precisely above the "S" on the horizon. In fact, the Sun appears slightly to the left, or east, of south. This

This figure-8 path of the Sun over the course of the year is explored further in the Analemma project.

small offset occurs for two reasons. First, Seattle is 2.3° west of the central meridian of its time zone (120° W, for the Pacific Time Zone), which places the Sun 2.3° east of due south at noon. Second, over the course of a year, the Sun at midday follows a long, narrow *"figure-8"* path in the sky called the **analemma**. In March, the Sun is near the eastern side of this figure-8 path, and will therefore appear slightly to the east of south at noon, even for someone exactly on the central meridian of a time zone. For someone away from the central meridian (e.g., someone in Seattle), the two effects add together.

3. To make sure that the time rate is set correctly, click on the **Time Interval** and select **300×**.
4. If Time Flow does not start immediately, click the **Flow Time Forward** button.

In fact, Venus actually does move relative to the Sun, due to its motion around the Sun. However, Venus's orbital motion is much slower than the diurnal motion of the sky due to the Earth's rotation and is not easy to discern in this view.

As time flows forward at 300 times its normal rate, notice that the Sun and Venus share the same motion across the sky. They remain separated by a fixed distance. This suggests that, in this relatively short time, we are seeing a constant motion of the entire sky as seen from the Earth, not just a motion of the Sun itself.

Compass directions are given along the horizon near the bottom of the screen. If S is centered, then SE and SW should be visible near the left and right sides of the screen, respectively. These compass directions show that motion toward the east is toward the left on the screen, and motion toward the west is toward the right.

Question 1:   Does the Sun's diurnal motion carry it eastward or westward on the screen?

Question 2:   Does this eastward or westward motion of the Sun correspond to motion toward the left or toward the right on the screen?

5. After observing this diurnal motion of the Sun and Venus, select **Go/Observing Projects/Diurnal A** from the menu to return to the initial configuration.

B. Motion of Other Objects in the Sky

During the daytime, air molecules in the Earth's atmosphere scatter sunlight making the sky appear bright. Short wavelengths are scattered more efficiently than long wavelengths, so blue (short-wavelength) light dominates this scattered light. The sky is so bright in the daytime that the only astronomical object we usually notice is the Sun. Occasionally, we can see the Moon, but it is fainter and less noticeable than the Sun. Of the planets, Venus is sometimes visible in the daytime as can be seen from this project, while Jupiter is occasionally visible if you know exactly where to look. All other astronomical objects are "lost" in the brightness of the blue sky. Nevertheless, even though we cannot see them, every astronomical object that is above the horizon is actually there in the sky during the daytime.

6. Select **View/Hide Daylight** from the menu to turn daylight off.

> **TIP:** You can also toggle daylight on and off with the keyboard shortcut Ctrl+D (Macintosh Cmd+D).

Turning daylight off eliminates the blue light, and you can see the Sun and the planets that are above the horizon at noon on March 1, against a background of stars. This is the view that you would have if the Earth had no atmosphere. The Sun, although very bright, would be just another object in the sky. Astronauts get a similar view of the sky from Earth's orbit or from the surface of the Moon.

If the motion of the Sun across the sky were due to the Sun actually moving, then the rest of the sky should remain stationary as the Sun moves past it; but if the apparent motion of the Sun is an illusion due entirely to our own motion, then our rotation around the Earth's axis should affect our view of all objects equally. Our rotation should make the entire sky appear to rotate around us.

7. Make sure that the Time Interval is correct by clicking on the **Time Interval** box and selecting **300×** and, when Time begins to advance, observe what happens.
8. Stop **Time Advance.**

Question 3:   What actually happens? Does the Sun move past the other objects in the sky while they remain at rest, or do all objects including the Sun appear to move equally?

C. Speed of the Sun's Diurnal Motion Across the Sky

The following question asks you to predict the speed of the Sun's diurnal motion; that is, how far the Sun moves across the sky each hour due to the Earth's rotation.

Question 4:   If the Earth rotates eastward through 360° (a full circle) in 24 hours, then through how many degrees and in which direction will the Sun appear to move in the sky in one hour? [*Hint:* If the Sun's apparent motion is due to the Earth's rotation, then the number of degrees through which the Sun moves each hour should be constant and equal to 360° divided by 24 hours, in degrees per hour.]

In the following steps you can verify your calculation by measuring the angle through which the Sun moves in 1 hour and compare it to your predicted value from Question 4.

9. Adjust the Time in the control panel to 10:00:00 A.M. Standard Time.
10. Change the **Time Step** in the control panel to **1 hour.**
11. Use **View/Hide Daylight** to remove the effect of daylight.
12. Measure the angular distance horizontally from the Sun to the left edge of the view. Enter this angle in Data Table 1 below.
13. Advance time by a single step of 1 hour.
14. Repeat the measurement of angular separation between the Sun and the eastern edge of the view and enter the angle in Data Table 1 below.
15. Repeat steps 13 to 14 to fill the Data Table.

IMPORTANT: When making these measurements, it is important that you do not inadvertently scroll the view. If you do, select **Edit/Undo Scroll** or use the keyboard shortcut Ctrl +Z (Cmd+Z on the Macintosh) and try the measurement again.

DATA TABLE 1

Measurement Number	Angle	First Differences of Angles
	Average Value	

HINT: It might be advisable to change each of these angles to arc seconds by multiplying arc minutes by 60 and degrees by 3600 and adding these values together with the arc second value.

The difference of two adjacent measurements (i.e., the "first differences" in the angles) will provide a measure of the Sun's motion in 1 hour. An average of these first differences in angle can be calculated to improve the accuracy of the overall measurement.

> **Question 5:** What is your measured value of the Sun's apparent motion in 1 hour? How does your measured value compare with the value you calculated in Question 4?

Your measured value should be close to your predicted value in Question 4. There will be small errors in these values because the Sun moves along a curved path, whereas the above measurements assumed that the Sun moved parallel to the horizon. Nevertheless, your measurements should agree reasonably with the calculated hourly rate of the Sun's motion.

D. Eastern Rising of the Sun in the Sky

It is interesting to explore the path of the Sun as it rises above the eastern horizon and to measure the angle that this path makes with the horizon from various latitudes. There is a great difference between sunrises (and in an equivalent manner, sunsets) when they are observed from different latitudes. For example, at the equator, the Sun appears rather suddenly and rises rapidly into the sky after only a short period of twilight. In contrast, at high latitudes the Sun moves more slowly across the horizon after a prolonged period of twilight. At very high latitudes during the summer, the Sun is always above the horizon.

In this section of the project, you will measure the angle between the Sun's track and the horizon, which we can call the "rising angle," and investigate the dependence of this angle upon the latitude of the observing site.

16. Select the view **Seattle** from **Go/Observing Projects/Diurnal Motion.**

The view shows the Sun rising in the East as seen from Seattle, USA, on March 19, 2005. This date, the first day of spring in the northern hemisphere (and the first day of fall or autumn in the southern hemisphere!), has been selected to place the Sun very close to due East at sunrise as a convenient reference point. The time step is initially set to 1 minute.

> **TIP:** An easy way to adjust the time and date fields is to click the desired field in the control panel and use the + and - keys on the keyboard to increment and decrement, respectively, the value of the field. Alternatively, use the shortcut keys as outlined in the User's Manual.

17. Use the single-step time controls to adjust the time in 1-minute intervals until the upper limb of the Sun is just visible above the horizon.
18. Change the **Time Step** to 1 hour.
19. Step the **Time Forward** 1 hour and note the direction in which the Sun has moved.

The Sun will have moved upward and, at most locations, to the side during this first hour after sunrise.

Question 6: In which direction has the Sun moved?

The geometry of the Sun's position is shown in the diagram below, with E its position at sunrise, X its position 1 hour after sunrise, and P the position directly below it on the horizon 1 hour after sunrise. (In practice, the track of the Sun across the sky will be slightly curved and the use of its position one hour after sunrise will underestimate the true rising angle, but this method is sufficiently accurate for our purposes.)

You can measure two angular separations on the sky and these can be assumed for this purpose to be distances. These angular separations are XE, between the Sun and due East, and XP, between the Sun and the horizon.

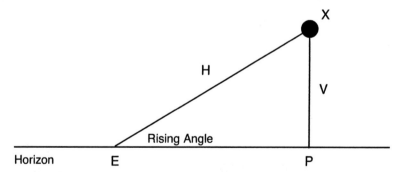

20. Use the angular measurement feature of the **Hand Tool** to measure the distance from the Sun to the East point of the horizon. Note this measurement in Data Table 2, below, under the heading H.
21. Next, measure the distance from the Sun to the point on the horizon directly below the Sun. Note this measurement in Data Table 2 under the heading V.
22. Open the **Info** pane and expand the **Location** layer. Note the latitude of the current viewing location and record it in the appropriate column of Data Table 2.

These angular separations, given in degrees, minutes, and seconds, need to be translated into degrees and fractions of a degree before you can use them as "distances." You can ignore the seconds because they are a small fraction of the final result. Divide the minutes by 60 and add this to the number of degrees (e.g., 34° 24' 45" would be 34 + 24/60 = 34.4°).

The ratio of these two "distances," $\frac{V}{H}$ is known as the trigonometric function Sine of the angle θ and is written as Sin θ.

23. Calculate the **ratio** $\frac{V}{H}$ and enter the value in Data Table 2.
24. Use the graph below to read off the value of the rising angle θ that corresponds to the calculated value of Sine θ. You can also use the **inverse sine** function on a scientific calculator or on the calculator software on your computer to calculate θ.

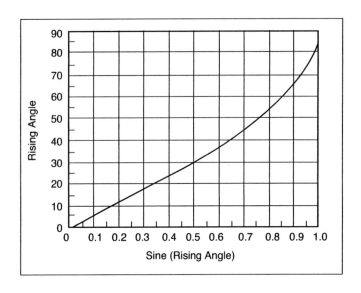

DATA TABLE 2

Location	Latitude	H	H in °	V	V in °	$\frac{V}{H}$ = Sine θ	θ
Seattle							
Honolulu							
Quito							
Dunedin							
Murmansk							
North Pole							

E. Dependence of the Sun's Rising Angle on the Observer's Latitude

It is interesting to explore the dependence of this "rising angle" of the Sun upon the latitude of the viewing location by repeating the above procedure for the other locations upon the Earth that are listed in Data Table 2.

25. For each location in Data Table 2, select the appropriately named view from **Go/Observing Projects/Diurnal Motion** and repeat the procedure outlined in steps 17 to 24 above.

The position and time of sunrise will vary somewhat between sites because of Time Zone differences. The situation at the North Pole is interesting. You will have to consider what is the "rising angle" if the Sun just skims across the horizon!

When you have entered data for each location in Data Table 2, try answering the following questions.

Question 7: At what angle from the horizon does the Sun rise for someone at 0° latitude?

Question 8: At what angle from the horizon does the Sun rise for someone at 90° latitude?

Question 9: How does the rising angle of the Sun change as you move from Quito, Ecuador (0° latitude), to Honolulu, then to Seattle, to Murmansk, and finally to the North Pole (90° latitude)?

Question 10: Based on your answers to the questions above, which ONE of the following statements do you think is correct?

a) The rising angle of the Sun at your location is equal to your latitude.

b) The rising angle of the Sun at your location is equal to 90° minus your latitude.

c) The rising angle of the Sun does not depend on your latitude.

Question 11: What is different between the sunrise in Dunedin, New Zealand compared with that in Seattle? [*Hint:* Think about the direction in which the Sun rises from the horizon.]

Question 12: Suppose that you are in Dunedin, New Zealand, the time is noon, and you are facing east, toward the point on the horizon where the Sun rises. Based on the direction toward which the Sun rises in Dunedin, in which direction (left or right) would you need to turn to face the Sun at noon in Dunedin? How does this compare to Seattle?

Question 13: Based on how the Sun rises in Dunedin, in which compass direction (north or south) would you need to face to see the Sun at noon? How does this compare to Seattle?

Question 14: Based on how the Sun rises in Dunedin, in which direction would the Sun be moving at noon (from left to right or from right to left)? How does this compare to what you saw in Seattle at noon?

26. Check your answer to Question 14 directly by selecting the file named Dunedin and adjusting the time to 12:00:00 PM (noon). **Drag the horizon around until you are facing north (N).** Set the **Time Step** to 1 minute, the Field of View to 100°, and check that the horizon is in the lower part of the view and relatively flat. Click **Flow Time Forward.**

TIP: Use the N key on the keyboard to move the view to the north.

Question 15: Toward which compass direction would the Sun be moving at noon in Dunedin (from east to west or from west to east)? Check the compass directions near the bottom of the view to find if you are right. How does this compare to what you would see in Seattle?

F. The Rising of Objects at Night

We have observed the rising of the Sun in both hemispheres. It is interesting to verify that other objects, when observed at night, also show the same motion and geometry as they pass over the horizon. Set up the following initial conditions and observe the eastern horizon as time advances.

27. Select the view **Seattle** under **Go/Observing Projects/Diurnal Motion.**
28. Reset the **Time** to **1:00:00** A.M.
29. Click **Flow Time Forward** and allow time to advance for a few hours.

As you observe the stars rise over the eastern horizon from Seattle, you see that they all rise at the same angle at which the Sun rose. This demonstrates once again that all objects undergo diurnal motion equally, during the day and at night.

G. The Setting of Objects in the West

We can also examine the geometry of sunset from Seattle.

30. Select **Seattle** from **Go/Observing Projects/Diurnal Motion** to return to the initial view from Seattle.
31. Set the **Time** to **4:00:00** P.M. and adjust the viewing direction to due **West**.
32. Click **Time Forward** to advance time to about **8:00:00** P.M.

You should see the Sun approach the horizon and set with the same angle to the horizon as it rose in the east. If you turn off daylight in the view you will see Venus (the very bright object below the Sun) setting with the same angle and speed as the Sun along with the stars of the springtime night sky. The fact that stars and constellations rise and set just like the Sun suggests that we are the ones who are moving, rather than the sky moving (as ancient astronomers believed). The apparent motion of the sky is an illusion caused by the Earth's rotation.

33. Close *Starry Night*™ without saving changes to the file.

H. Conclusions

In this project you have demonstrated that the sky appears to move from east to west because of Earth's spin on its axis, carrying observers toward the east. You have also evaluated the relationship between the observer's latitude and the angle between the horizon and the object's track at sunrise. You have also observed the difference in sunrises between the two hemispheres and unusual sunrises at the North Pole.

THE EARTH'S ORBITAL MOTION 4

THE SUN'S APPARENT MOTION ACROSS THE SKY

Our view of the night sky is constantly changing. Two of the most evident changes are the apparent motion of the stars and constellations toward the west over the course of a single night and the much slower shift of stars and constellations toward the west over the course of a year when viewed at the same time each night. The **first** of these changes is caused by the Earth's rotation. This makes the sky appear to rotate around the Earth's polar axis once every day, or at a rate of 15° per hour and makes the Sun, Moon, and stars rise in the east and set in the west each day. The **second** change is caused by the Earth's orbital motion around the Sun. Because the Earth makes a full orbit of 360° in about 365 days, our view of the universe at any given time of the night changes by about 1° per day. The combined result of these two changes is that, in terms of our Sun-based time, each star rises about 4 minutes earlier each night.

If we could see the stars in the daytime, the most obvious effect of our orbital motion around the Sun would be an apparent motion of the Sun past the background stars. In real life this motion is hidden from us by the blue sky of daylight, which is the result of scattering of sunlight by the Earth's atmosphere. In the virtual sky of *Starry Night™*, we can eliminate the blue sky of daylight to reveal the Sun's apparent motion against the background sky.

The plane of the Earth's orbit, which contains the Sun, is known as the **ecliptic plane,** and the projection of this plane onto the sky forms a line or circle around the sky called the

Section 2-5 and Figure 2-12 in Freedman and Kaufmann, *Universe,* 7th Ed., discusses the seasons. Figure 2-14 illustrates the ecliptic plane.

ecliptic. This path passes through a limited set of constellations known as the **Zodiac.** The zodiac is a band of constellations through which most of the brightest planets and the Moon appear to move as they orbit close to the ecliptic plane. This region of our sky played an important role in early astrology because the positions of the planets were used to predict the future and these zodiacal constellations or "signs" are still used in modern astrology.

This observing project will demonstrate the daily motion of the Sun past the background stars of the Zodiac as it traces the ecliptic around the sky. It will also allow you to measure the apparent speed of the Sun against this background and to show that this speed varies over the year because of the Earth's elliptical motion.

A. The Sun's Apparent Motion Across our Sky

1. Launch *Starry Night™* and open the view **Orbit A** under **Go/Observing Projects/Earth's Orbital Motion.**

The view shows the sky looking south from near Chicago at noon on December 21, 2006. Daylight has been turned off to reveal the background stars. The Sun is near the center of the view and labeled. The time step is set to 1 sidereal day. The sidereal day is the true rotation period of the Earth with respect to the background stars. Thus, in one sidereal day, the background sky returns to exactly the same position in the view and the stars appear to remain stationary as we advance time. The motion of the Sun against these background stars is then much easier to see.

   2. Single-step Time Forward by 1 sidereal day at a time.

> **TIP:** You can use the **U** key as a shortcut to advance one time step and **Shift+U** to move back one time step.

The Sun appears to move day by day because of our motion on an orbiting Earth. As you advance time by 1 sidereal day, you will notice that the time displayed in the Time box, which is Solar Time, actually advances by less than 1 day. This is because one sidereal day is less than 1 solar day by 3 minutes, 56 seconds. In other words, 1 sidereal day is equal to 23 hours, 56 minutes, 4 seconds in Solar Time.

   Question 1:   In which direction does the Sun appear to move in our sky in 1 sidereal day?

   3. Reopen the view **Orbit A** from **Go/Observing Projects/Earth's Orbital Motion** to return to the initial view and then run **Time Forward** to observe the motion of the Sun across the sky. You may want to repeat this step several times.

You can see that the Sun appears to move smoothly with respect to the background stars as time changes in steps of one sidereal day. Although the background stars remain fixed as you advance time, you may notice several starlike objects that move across the sky at a different rate to that of the Sun. These are, of course, **planets.** It is easy to see why these objects were called "wandering stars" by early astronomers.

   4. Re-open the view **Orbit A** from **Go/Observing Projects/Earth's Orbital Motion** to return to the initial settings.
   5. To identify the planets that accompany the **Sun** in this view, open the **View Options** pane, expand the **Solar System** layer, and click the box to the right of **Planets/Moons.**
   6. Use the **Single Step Forward** button to watch the planets, particularly Mercury and Venus, move along their orbits against the stationary background stars.

B. Measurement of the Sun's Apparent Speed Across the Sky

The apparent speed of the Sun across the sky can be measured and compared to the predicted speed by using the angular measurement feature of the Hand Tool. You can do this by choosing an appropriate reference star close to, but behind, the Sun on its path. You can then measure the successive angular separations of the Sun from this star as you advance time. The differences between these successive angles will provide a measure of the Sun's speed in degrees per selected time interval.

   7. Open the view **Orbit B** under **Go/Observing Projects/Earth's Orbital Motion.**

The view is the same as in the previous section except that the time step is now 10 sidereal days and the ecliptic is shown. The star *Lambda Librae,* which lies on the ecliptic, is labeled. We will use this star as a reference point against which to measure the motion of the Sun over time.

8. Place the **Hand Tool** over the Sun and measure the angular separation from the Sun to *Lambda Librae.* Note this separation and the appropriate date in Data Table 1 below, under Data Number A.
9. **Single-step Time Forward** by 1 step of 10 sidereal days, repeat the angular separation measurement, and record this and the date under Data Number B.
10. Repeat this procedure for two more 10 sidereal day steps and enter the dates and angular separations under Data Numbers C and D.
11. The angular motion of the Sun in 10 sidereal days can be determined by taking the difference between adjacent measurements of angular separation. Enter these differences in the Data Table.

DATA TABLE 1

Data Number	A	B	C	D
Date				
Angular separation				
Difference in angular separation (e.g., B-A)				

Question 2: What is the average speed of the Sun in degrees per 10 days around the beginning of January?

Section 4–4 in Freedman and Kaufmann, *Universe,* 7th Ed., discusses motion in elliptical orbits.

Although the Sun appears to move smoothly across our sky with evenly spaced daily steps, its speed varies because the Earth moves in an elliptical orbit, moving faster at some times and slower at others. Because we are on the Earth, our variable speed makes the Sun's apparent speed across our sky appear to vary.

The above measurements were made close to the date when the Earth was moving at its fastest in its orbit, around January 4. On this date, the apparent motion of the Sun from the Earth appears to be fastest. It is possible to compare this speed with that when the Earth is moving at its slowest six months later, in early July, by repeating the above series of measurements in late June and early July.

12. Open the view **Orbit B2** under **Go/Observing Projects/Earth's Orbital Motion.**

The view shows the Sun and the ecliptic as seen at noon from near Chicago on June 21, 2007. The horizon is not in this field of view in midsummer from Chicago because of the high elevation of the Sun at this time. A reference star on the ecliptic, 13 *Tauri,* is labeled. The time step has been set to 10 sidereal days.

13. Measure the angular separation from the Sun to the reference star and record the date and resulting measurement in Data Table 2 below.
14. **Single-step Time Forward** in 10 sidereal day steps, and repeat the angular separation measurement recording the results in the Data Table.

DATA TABLE 2

Data Number	A	B	C	D
Date				
Angular separation				
Difference in angular separation (e.g., B-A)				

15. Calculate the differences between successive angular separations to determine the speed of the Sun in degrees per 10-day interval.

Question 3: What is the speed of the Sun in degrees per 10 days around the beginning of July?

Question 4: In which of the months, January or July, is the speed of the Earth the highest?

C. Motion of the Sky in Solar Time

In the steps above, you advanced the time in sidereal days in order to keep the stars fixed in the sky. This allowed you to watch the apparent motion of the Sun relative to the stars. It is interesting to examine the change in appearance of the sky as time changes in units of solar days rather than sidereal days. In this case, the stars will move across your sky while the Sun remains at approximately the same azimuth, in the present case almost due south. The Sun will move through a range of elevation angles because of the tilt of the Earth's spin axis to its orbital plane. This change of elevation angle is important in defining the seasons upon the Earth.

16. Open the view **Orbit A** under **Go/Observing Projects/Earth's Orbital Motion.**
17. Display the local meridian at Chicago by opening the **View Options** pane, expanding the **Guides** layer, and clicking on **Local Meridian.**

You will note that the Sun is not on the meridian at midday because of Chicago's position in the Time Zone.

> **Time Zones** are discussed in Section 2-7 and displayed in Figure 2-21 of Freedman and Kaufmann, *Universe*, 7th Ed.

18. Adjust the **Time** to place the Sun on the local meridian on this date at the winter solstice.
19. Change the **Time Step** to 1 day.
20. Drag the sky downward until the Sun is very near the bottom of the view.
21. Run **Time Forward** and watch the motion of the sky and the Sun in solar days.

Careful observation of the Sun over a year will reveal small deviations from due south in the Chicago sky, that is, a deviation from the meridian as the Sun's position moves in the north-south direction because of the tilt of the Earth's axis from the perpendicular to the Earth's orbit. This deviation is another representation of the variation in speed of the Earth in its orbit and is illustrated in the Analemma project.

One further interesting exercise is to track the Sun as it passes along the ecliptic and through the familiar constellations of the Zodiac.

22. Open the view **Orbit C** under **Go/Observing Projects/Earth's Orbital Motion.**

The view is essentially the same as that of Orbit A except that the constellations of the zodiac are shown and labeled, the ecliptic is shown, the time step is 1 day rather than 1 sidereal day, and the view is centered upon the Sun in order to prevent it from moving off the screen.

23. Run **Time Forward** to show the progression of the Sun across the familiar constellations of the Zodiac. (For interest, you might like to stop this motion briefly on your birthday and check whether the Sun is within your astrological birth sign on the Zodiac.)

You will see the Sun passing along the ecliptic but you will also note that the planets stay within the band of stars making up the Zodiac. The appearance of the sky in this view is equivalent to the view from the moving outside edge of a merry-go-round, watching its center. The Sun represents this center in our solar system. Because you are watching the center, it remains fixed in your view while the surroundings beyond the merry-go-round revolve continuously around you. For the same reason, the constellations that we see at night shift slowly toward the west in a 1-year cycle when we watch the sky from day to day (or night to night). Being on the Earth as it orbits the Sun is like being on a gigantic merry-go-round, taking 1 full year to make a full circle.

> The **Zodiac** is discussed in Section 2-6 of Freedman and Kaufmann, *Universe*, 7th Ed.

24. Exit *Starry Night*™ without saving changes to the files.

D. Conclusions

This project has shown you the apparent motion of the Sun through our sky in a way that is impossible to demonstrate in the real world because of the presence of scattered sunlight in the daytime. Furthermore, we have been able to measure the variation in speed of the Sun across the sky caused by the variation of the speed of our observing platform, the Earth, as it moves in its elliptical orbit.

THE SEASONS AND THE LENGTH OF THE DAY

5

In temperate latitudes, there is a distinct division of the year into the seasons of spring, summer, autumn, and winter. Summers are warm and winters are cool. These seasons are caused by the tilt of the Earth's axis of rotation relative to the direction to the Sun.

In June, the North Pole is tilted toward the Sun and the South Pole is tilted away, giving more direct sunlight and warmer temperatures to northern latitudes and less direct sunlight and cooler temperatures to southern latitudes. Thus, it is summer in the northern hemisphere and winter in the southern hemisphere.

In December, the North Pole is tilted away from the Sun and receives less direct sunlight, while the South Pole is tilted toward the Sun and receives more direct sunlight. At this time the northern latitudes are cooler and the southern latitudes are warmer; i.e., it is winter in the northern hemisphere and summer in the southern hemisphere.

> Section 2–5 and Figures 2–12 and 2–13 in Freedman and Kaufmann, *Universe*, 7th Ed., describe the seasons and their relationship to the tilt of the Earth's axis.

The Earth's rotation axis points almost exactly at the North Star (Polaris), and this alignment does not change with the seasons. (There is a very slow, cyclic change over a period of 26,000 years, in a process called precession, but the change over 1 year is very small and does not affect the seasons.)

In this observing project, we will investigate how the times and locations of sunrise and sunset change with the seasons and how these seasonal changes vary with latitude on the Earth.

A. The Position of Earth's Axis Relative to the Sun

In this section, you will have the opportunity to observe the Earth from several unique perspectives that demonstrate the cause of the seasons.

> 1. Launch *Starry Night*™.
> 2. Select **View over S Pole** from **Go/Observing Projects/Seasons**.

The view is centered on the Earth from a position about 20,600 km above the South Pole on March 21, 2007, at 00:06:40 UT. In other words, the direction of the view is along the polar axis of the Earth toward the North Celestial Pole. The celestial grid, the coordinate system based on the rotation axis of Earth is shown. The North Ecliptic Pole, the position in space perpendicular to the plane of Earth's orbit around the Sun is labeled and appears near the top of the view.

3. Use the **M key** on the keyboard to advance time in increments of 1 calendar month and observe the position of the Ecliptic pole. Advance time through a full calendar year to complete your observations.

As you progress through the year, you will note that the South Pole is in darkness for 6 months of the year, from late March to late September, and illuminated by the Sun for the other 6 months, from late September to late March. This region of the Earth thus suffers extreme seasonal variations as a consequence of the tilt of its axis to the perpendicular to the ecliptic plane containing the Sun, represented by the North Ecliptic Pole.

4. Return to the initial setup by selecting **View over S Pole** from **Go/Observing Projects/Seasons** and increase the elevation to about 0.001 AU.
5. With the date reset to **March 21**, use the **Hand Tool** to measure the angular separation between the Earth and the North Ecliptic Pole to the nearest $1/2°$. Repeat the measurement for June 21, September 21, and December 21.

Question 1:   What is the angular separation between the polar axis of the Earth (NCP) and the Ecliptic Pole to the nearest $1/2°$?

Question 2:   Does the angular separation between the NCP and the Ecliptic Pole vary significantly over the course of a year?

6. Select **View over S Pole** in **Go/Observing Projects/Seasons** again and click the **Location** field in the control panel. Choose **Earth's Center** from the list.
7. In the **View Options** pane, expand the **Guides** layer and activate **Celestial Poles** from the menu.

This view is almost identical to the original view except that you are now at the center of a transparent Earth. You are still looking directly north along the polar axis of the Earth.

8. Advance time through a full year, 1 calendar month at a time.

Question 3:   Over the course of 1 year, does the rotation axis of Earth (equivalent to the North Celestial Pole) move against the background stars?

Question 4:   Does the position of the Ecliptic Pole move with respect to the background stars over the course of a year?

Question 5:   How many months elapse before the Ecliptic Pole returns to its original position in the view?

The sequence shows the view along Earth's rotation axis from a coordinate frame tied to the Earth. The sequence shows that this rotation axis points in a relatively fixed direction with respect to the stars throughout the year. The ecliptic pole, the axis of the plane of the Earth's orbit about the Sun, moves in a circular pattern along with the background stars from this perspective, completing one full cycle in one year.

9. Select **View from Sun Center** under **Go/Observing Projects/Seasons.**

The time and date are the same as in the previous sequences. The view is now from the center of a transparent Sun, looking toward the Earth along its orbital plane. The tilt of the rotation axis of Earth relative to its orbital plane is indicated clearly by the angle between the Celestial Equator, which is the plane of the Earth's equator at right angles to its rotation axis, and the ecliptic plane, which is the plane of Earth's orbit.

10. Advance time through a full year in intervals of 1 calendar month.

Question 6: As seen from the Sun, which of the following best describes the pattern traced out by the polar axis of the Earth over the course of 1 year?
a) Cylinder
b) Cone
c) Line
d) Stationary point

Question 7: During which month is the North Pole of the Earth tilted toward the Sun?

Question 8: During which month is the South Pole of the Earth tilted toward the Sun?

B. The Position and Time of Sunrise Through the Year

In this part of the project, you will see how the location of sunrise on the horizon changes throughout the year at one location in the northern hemisphere and how this variation is matched by changes in the time of sunrise.

11. Select **Sunrise** from **Go/Observing Projects/Seasons.**
12. Open **File/Preferences/Cursor Tracking (HUD)** and select only **Azimuth** and **Name** from the **Show** list.

Starry Night™ shows the view looking east at sunrise from New York on March 21, 2007. In the following sequence you will observe the position and time of sunrise through the year, recording the results of your observations in Data Table 1 below.

13. Click the **seconds field** of the Time in the control panel and use the + and – keys to adjust the time so that the upper edge of the Sun is just visible above the horizon. Enter the Time of Sunrise in Data Table 1.
14. Position the **Hand Tool** over the Sun and note the azimuth of the Sun from the HUD. Record this value under the Azimuth column in Data Table 1. Subtract this value from 90° (due east) and enter the result, with the proper sign, into the Data Table. The result of this calculation will be positive when the Sun rises to the north of east and negative when it rises to the south of east.

Azimuth is measured from north (0°) through east (90°), south (180°), and west (270°).

DATA TABLE 1

Date	Time of Sunrise	Azimuth	90° − Azimuth
3/21/2007			
4/21/2007			
5/21/2007			
6/21/2007			
7/21/2007			
8/21/2007			
9/21/2007			
10/21/2007			
11/21/2007			
12/21/2007			
1/21/2008			
2/21/2008			
3/21/2008			

15. Increment the date by 1 calendar month.
16. Use the + and − keys to adjust the minutes and seconds of the Time in the control panel until the upper edge of the Sun is just visible above the horizon. Record the time of sunrise in Data Table 1.
17. Position the **Hand Tool** over the Sun and note its azimuth from the HUD in Data Table 1.
18. Calculate the difference between the Sun's azimuth and due east and record the result of the calculation in Data Table 1.
19. To enter the data for the remaining dates, repeat instructions 15 to 18 for each month.

Note: From July to December, the Sun will be below the horizon after you change the month so you will have to advance time (use the + key) until the Sun rises above the horizon; then adjust the time backward until its upper limb is just visible above the horizon.

Now you can use your recorded measurements to plot the sunrise position (represented by [90° − Azimuth] in the data table above) graphically against the date on a sheet of graph paper. Mark off the time horizontally along the bottom (the "x-axis"), and the angle north or south of east vertically along the left side (the "y-axis").

A quick way to number the horizontal x-axis that is accurate enough to see the trend is to mark off 12 equal months, starting with March 21 (Spring Equinox), then April 21, etc., to February 21, and finish with March 21 (i.e., start and end with March 21). This method is not precise because it assumes that all months have the same length, which is not the case, but is sufficient for the present purpose.

For the vertical y-axis, mark off eight equal intervals from 40° south of east (you can label this −40°) through 0° to 40° north of east (or +40°). It will make plotting easier if you can make each interval 10 (or a multiple of 10) units in size on the graph paper; then, for example, 28° would be on the eighth line above the 20° line.

Plot a second graph, this time showing the sunrise time against the date in the year. You might use a scale from 4:00 A.M. to 8:00 A.M. along the vertical y-axis, or a larger range than this if you find it is more convenient.

Question 9: What is the shape of the graph of sunrise position as a function of time of the year (e.g., straight line, sinusoidal [i.e., S-wave] curve, etc.)?

Question 10: In which two months of the year does the Sunrise occur exactly (or almost exactly) at the due east position?

Question 11: In which months does it rise to the north of east? During these months, does it rise earlier or later than it does on March 21?

Question 12: In which months does it rise to the south of east? During these months, does it rise earlier or later than it does on March 21?

Question 13: In which month does the Sun rise farthest north? In which month does it rise farthest south? Which of these months corresponds to the latest sunrise? Earliest sunrise?

Question 14: During which months is the Sun moving northward (sunrise farther north than it was the previous month)? During which months is it moving southward?

Question 15: What happens to the time of sunrise as the Sun moves northward? Southward?

Question 16: Based on your graph, in which months is the sunrise position changing most rapidly (greatest change per day)?

Question 17: At which point on the horizon does the Sun rise when the time of sunrise is changing most rapidly? (Compare your sunrise time graph to your sunrise position graph to answer this question.)

You can now use your graphs to tell where on the New York horizon the Sun will rise on any day of the year, and estimate the time at which it will rise. Remember that the graphs are plotted for the 21st of each month.

Question 18: According to your graphs, what is the time of sunrise in New York for November 5; i.e., one-half month after October 21? How far south of east will it rise?

C. Change in the Length of the Day Through the Year

Having observed sunrise throughout the year from the latitude of New York, you can now make observations of sunset from the same location in this section of the project.

20. Select **Sunset** under **Go/Observing Projects/Seasons.**

The view is toward the west near sunset on March 21, 2007, from New York City.

21. Adjust the minutes and seconds fields of the time in the control panel to adjust the time so that the upper edge of the sun is about to disappear below the horizon. Note the time of sunset in Data Table 2.
22. Change the date in the control panel to the other dates in Data Table 2 and repeat the previous step, recording the time of sunset for each date.

DATA TABLE 2

Date	Time of Sunrise	Time of Sunset	Length of Day (h/m/s)	Length of Night (h/m/s)
3/21/2007				
6/21/2007				
9/21/2007				
12/21/2007				

In Data Table 2, use the sunrise times that you measured in Part B with the sunset times you measured in this section of the project to calculate the length of the day in New York City for all four dates. Also calculate the length of the night, which will be equal to 24 hours minus the length of the day.

> **Question 19:** What approximate relationship do you notice between the lengths of the day and night for June 21 and December 21? (For instance, how does the length of the day on June 21 compare to the length of the night on December 21?)
>
> **Question 20:** Based on your values in Tables 1 and 2, what happens to the length of the day as the rising or setting point of the Sun moves northward along the horizon? Southward?
>
> **Question 21:** We have kept the time set to Standard Time throughout this exercise, but in fact most areas use Daylight Saving Time (or Daylight Time) between April and September. If Daylight Time were in effect, would it have any effect on the length of the day or night? (Remember that we set our clocks one hour ahead when daylight time begins, so sunrise AND sunset take place 1 hour later than they would if we had left the clocks at Standard Time.) What then does the phrase "Daylight Saving" mean?

D. Length of the Day at the Equator

As you have seen, summer brings longer days and shorter nights to mid-northern latitudes. Conversely, in winter, the days are shorter and nights longer. The length of the day over the course of the seasons depends, however, upon one's latitude on Earth.

23. Select the view **Length of Day-Equator** from **Go/Observing Projects/Seasons**.

The view is centered on the Sun on March 21, 2007, and the field of view decreased to provide a close-up of the Sun at around the time of sunrise as seen from the equator on the Earth.

24. Adjust the time so that the upper limb of the Sun is just above the eastern horizon. Note the time of sunrise for this latitude and record it in Data Table 3.
25. Note the azimuth of the Sun at sunrise. Subtract the azimuth from 90° and record the result of this calculation including the sign under the column 90° – azimuth in Data Table 3. This gives you the sunrise position relative to east with positive values indicating a position north of east and negative values a position south of east.

> **NOTE:** Use the HUD to determine the azimuth of the Sun at sunrise.

26. Change the A.M. field of the time in the control panel to P.M. Because the view is locked on the Sun, the view changes to the west near the time of sunset. Adjust the time in the control panel to sunset, the point at which the last sliver of the Sun is about to disappear below the horizon. Record the time of sunset in Data Table 3.
27. Find the time of sunrise, azimuth of the Sun at sunrise, and the time of sunset for each of the other dates listed in Data Table 3 using the techniques of the previous three steps. Record these times in the Data Table.
28. Calculate the length of the day and of the night at the equator for each of the dates.

DATA TABLE 3

Date	Sunrise Time	90° – Azimuth	Sunset Time	Length of Day
3/21/2007				
6/21/2007				
9/21/2007				
12/21/2007				

In the next sequence, you will observe sunrise and sunset from the North Pole.

29. Select the view **Length of Day-North Pole** from **Go/Observing Projects/Seasons.**

The view is centered upon the Sun. From the North Pole on March 15, 2007, at 6:00:00 A.M., the Sun is still below the eastern horizon.

30. With the time step set to one hour, start the time flow forward, stopping time flow as soon as the Sun rises above the horizon. Note the date on which the Sun rises as seen from the North Pole during 2007.
31. Change the time step to 4 hours and with the view locked on the Sun, start the flow of time forward. Stop time flow when the Sun nears the horizon again and use single steps to determine the date of sunset in 2007 as seen from the North Pole.

Question 22: How long is the day as seen from the North Pole?

E. Alignment of Box Hill Tunnel and Brunel's Birthday

As an extension to this project and as an interesting historical supplement, you can solve this puzzle.

In the 1800s, the great engineer and ship builder Isambard Kingdom Brunel completed many remarkable projects including construction of the giant ship the "Great Eastern," the largest ship in the world for 40 years. Driven by screw and paddle wheels, the Great Eastern achieved fame by laying the first transatlantic telephone cable.

A picture of this remarkable tunnel, as well as information on Brunel, can be found at: www.spartacus.schoolnet.co.uk Click on Railways: 1780-1900, then Isambard Kingdom Brunel, then Box Tunnel.

In 1841, Brunel completed a 2-mile-long railway tunnel through Box Hill in England for the Bristol-to-London trains that was remarkable for being so straight that the sun would shine along its complete length at sunrise on 2 days of the year. Furthermore, he aligned it such that one of these days was his birthday!

With the fact that this tunnel is aligned at an azimuth of 78.5° and the knowledge that Brunel was born in the spring of the year 1806, you can use *Starry Night*™ to determine his birthday.

32. Select the view **Box Hill** under **Go/Observing Projects/Seasons.**
33. Adjust the **Time Forward** or **Backward** on January 1, 1841, until the Sun is just above the horizon.
34. Use the **Hand Tool** to find the Sun's azimuth from the HUD.
35. Adjust the date, initially in 1-month steps, and repeat the previous two steps to adjust the time to sunrise and determine the Sun's azimuth. When you get close to the correct date (i.e., when the azimuth of the Sun at sunrise approaches 78.5°), it is useful to zoom in to a field of view of 30° and change the time step to 1 minute, making sure that the Sun and the eastern horizon are on the screen. In this way, you can place the full Sun just above the horizon.
36. Adjust the Date until the Sun rises at an azimuth of 78.5° on the horizon at Box Hill.

For interest, you can run the simulation ahead a few months and repeat the above procedure to discover the other day in the year when the direction toward sunrise would coincide with the tunnel direction. In practice, the floor of the tunnel slopes at about one-half degree to the horizontal and this would have to be taken into account in order to obtain an accurate result. Of course, there will also be 2 days in the year when the sunset sun would shine along the tunnel in the opposite direction, this time from 11.5° south of due west on the horizon (azimuth 258.5°). You can use the above method to determine these two dates. Nevertheless, it would have been disconcerting, at the very least, for railway engine drivers to find the tunnel flooded with light on these specific times every year!

Question 23:   What was Isambard Kingdom Brunel's birthday?

F. Conclusions

You have explored the times of sunrise and sunset and the relative lengths of day and night in winter and summer. The tilt of the Earth's axis is the main contributor to the seasonal effects upon midlatitudes, where the changing inclination of sunlight to the Earth's surface changes the efficiency of heating. Nevertheless, the length of day also affects the seasons by providing longer growing times in summer.

You also used *Starry Night*™ to determine the birthday of one of the great engineers of the nineteenth century, who obviously had a sense of humor (or of history and his place in it!) in aligning a railway tunnel toward sunrise on his birthday.

PRECESSION AND NUTATION OF THE EARTH

6

As the Earth orbits the Sun in the ecliptic plane, it also rotates in space with a period of one sidereal day around an axis that is inclined to this orbital plane. This angle of inclination varies over very long periods but is about $23\,^1/_2°$ at the present time. If the Earth's shape were a perfect sphere, this spin would be undisturbed and the direction of the spin axis would remain fixed in space. However, the Earth's rapid rotation, particularly during its formative stages, has distorted it into a flattened shape known as an oblate spheroid in which its equatorial diameter is now 43 km longer than its polar diameter. In addition to the direct gravitational forces that act between the centers of the Earth, the Moon, and the Sun to maintain their respective orbital motions, this slightly distorted Earth now becomes subject to the influence of differential gravitational forces from the Moon and the Sun.

> These differential forces also cause the tides in the Earth's oceans by acting with slightly different strengths on equivalent masses on either side of Earth. These tidal effects are discussed in Section 4–8 of Freedman and Kaufmann, *Universe*, 7th Ed.

> Precession is discussed in Section 2–6 of Freedman and Kaufmann, *Universe*, 7th Ed.

These differential forces act upon Earth's equatorial bulge to cause its spin axis to move in a conical motion called **precession.** Because the Moon and Sun remain close to the ecliptic plane, they generate an average force on the bulge at an angle of $23\,^1/_2°$ to the equatorial plane. This disturbs the simple spinning motion of the Earth.

This slow coning precessional motion of the axis is equivalent to that of a child's spinning top. When a spinning top is placed upon the ground with its axis inclined to the vertical, there are out-of-balance forces acting upon the top that are attempting to cause it to fall over. Indeed, if it were not spinning, the top would fall over under gravity. In both cases—spinning top and rotating Earth—out-of-balance forces cause classical precession of the spin axis. The actual dynamics are relatively simple but require knowledge of the physics of rotational motion that is beyond the scope of this book.

The consequences of this precessional motion upon life on Earth are very small and affect almost no one. However, the equatorial plane and the spin axis of Earth define **right ascension** and **declination,** the celestial coordinates used by astronomers to define positions of objects in our sky. Precession therefore causes the right ascension and declination of each star to change slowly and continuously over time as the direction of the Earth's spin axis changes. Star atlases have to be specifically labeled with the epoch for which they are applicable. Those in use today have been drawn for the epoch 2000.0. In order to point a telescope accurately at any object in the sky whose right ascension and declination are known from this atlas, corrections have to be made night-by-night to allow for this slow drift. This adjustment is known as *precessing the coordinates* of an object and is always done before pointing the telescope toward an object. In modern telescope systems, this adjustment is carried out automatically by the telescope control system.

In this project, we will display on the sky and measure the precessional motion of the northern extension of the spin axis of the Earth, the North Celestial Pole or NCP, by compressing time so that we can "observe" the motions of the sky over very long periods. We will also examine and measure

Figure 2-14 of Freedman and Kaufmann, *Universe*, 7th Ed., illustrates the vernal equinox.

the motion of the **vernal equinox** across the sky because of precession. The vernal equinox is one of the two points at which the Earth's orbital plane intersects its equatorial plane and is used as the zero point of right ascension in the celestial coordinate system. The Sun passes through the vernal equinox on the first day of spring every year, hence its name. In the final section of the project, we will observe and measure a tiny additional wobble of the spin axis of Earth, known as **nutation**, which is superimposed upon the smooth and long-term precession.

A. Observation of Precessional Motion

We will use *Starry Night*™ to view the northern polar region of the sky in order to observe the motion of the spin axis of Earth against the background of the stars. We will then follow the motion of the North Celestial Pole against the background stars as time is advanced rapidly. The NCP is easily found at the present epoch because it is close to a bright "marker star," Polaris, our so-called Pole Star.

1. Launch *Starry Night*™.
2. Select **File/Preferences** from the menu. In the Preferences dialog, open the **Cursor Tracking (HUD)** page and choose only **Name** and **Object type** from the **Show** list.
3. Select the view **Precession-A** under **Go/Observing Projects/Precession**.

The view shows the northern sky as seen from Toronto, Canada, on December 14, 2004 at 9 P.M. local standard time. The North Celestial Pole (NCP) is near the center of the view, marked by a red cross. Polaris is visible just below the NCP. This wide-angle view will remain fixed upon the NCP at the center of the screen as time flows, and the stars will appear to move. In reality, the stars represent a fixed background against which the Earth's coordinate system and the NCP will move.

4. Run **Time Forward** and **Time Backward** to follow the track of the NCP across the sky between the **Date** limits, 4713 B.C. to 9999 A.D.

You will see that, as a result of precession, the NCP appears to move in a circle around a fixed position among the stars.

5. Select the view **Precession-A2** under **Go/Observing Projects/Precession**.

The view is identical to that of the previous sequence except that several stars as well as the north ecliptic pole are labeled. The line from the Earth's center to the north ecliptic pole is perpendicular to the ecliptic plane. This line is the axis of the cone through which the spin axis of Earth moves under precession, the end of which is the fixed point you observed in the previous sequence. The cone angle is about $23\frac{1}{2}°$. As you change the date, note how fortunate we are at the present time in history in the Northern Hemisphere to have a star as bright as Polaris, our "Pole Star," so close to the NCP.

6. Run **Time Forward** and **Time Backward** to follow the track of the NCP across the sky between the **Date** limits, 4713 B.C. to 9999 A.D., and note that the north ecliptic pole marks the fixed position around which the stars appear to rotate.

It is interesting to use this time compression to investigate the position of the NCP in history and in the future and answer the following series of questions.

7. Manipulate the zoom and time controls in the control panel (including adjusting the time step interval) and use the angular measurement feature of the **Hand Tool** to help you to answer the following questions. (*Hint:* You might want to Center and lock on to the star in question to make some of these measurements.)

Question 1: When is the NCP closest to Polaris, our Pole Star?

Question 2: How close in angle will Polaris be to the NCP on this date?

Question 3: Does the NCP approach any stars of equivalent brightness in the time period 4713 B.C. to 9999 A.D.? If so, which stars are these?

Question 4: What is the year of closest approach of the NCP to the star Thuban and what is the angular spacing between NCP and Thuban on this date? To which of these two stars, Polaris or Thuban, did (or will) the NCP approach closest in this precessional motion?

8. Select the view **Precession-A2** again from Go/Observing Projects/Precession and run **Time Forward** until it reaches its maximum date.

The maximum date available in *Starry Night*™ does not allow one to see the close approach of the NCP to the very bright star Vega in the constellation, Lyra, but one can see in 9999 A.D. that it is moving toward this star.

9. Because you know that the radius of the circle that the NCP makes around the North Ecliptic Pole is about 23.45°, you can use the angular measurement feature of the **Hand Tool** to estimate how close the NCP will approach Vega at some time in the future by measuring the angle between Vega and the ecliptic pole.

Question 5: How close will the NCP approach Vega during its precessional motion, at some time in the future?

It is possible to examine the constellations through which the NCP passes on this extended passage across the sky by toggling on the constellations.

10. Select the view **Precession-A** under **Go/Observing Projects/Precession.**
11. Open the **View Options** pane and select **Boundaries** and **Labels** options under the **Constellations** layer.

Question 6: In which constellation is the NCP located today?

Question 7: In which constellation was the NCP at the time of the early construction of Stonehenge in southern England, in about 2500 B.C?

B. Measurement of Precessional Motion of the North Celestial Pole

We can measure the precessional motion of the spin axis of Earth easily by finding a time, in history or in the future, when this axis is pointing toward a star in our sky. We can then use this star as a reference point and measure the angle moved by the end of the axis, the NCP, after a specified period. An example of this type of close approach occurred in August 1449 A.D. when the NCP was close to the star designated **HIP59415.**

12. Select the view **Precession-B** under **Go/Observing Projects/Precession.**

The view of a 24 arc minute field of view on August 1, 1449 A.D., shows the NCP and a labeled reference star, **HIP59415.** (The time step is set to an interval of 19 years, for reasons that will become clear in Section D below.)

13. Use the **Hand Tool** to drag the view so that the bright red line appearing diagonally across the screen (the celestial meridian) becomes horizontal and the NCP is in the center of the view.

The date has been chosen to place the NCP as close as possible to the above reference star, **HIP59415.** Because the view is locked on to the NCP, a time step forward or backward by 19 years will move this reference star from the center toward the edge of the screen. By taking the star's position of closest approach to the NCP as a reference position, you can measure the angular separation between the star and its reference position just above the NCP on these two dates, 19 years either side of 1449 A.D., and then average these values to provide the angular movement of the NCP across our sky in 19 years.

With a little simple trigonometry, you can then determine the change in the orientation of the Earth's spin axis per year and calculate the time taken for the axis to complete one full cycle of 360°.

14. **Step Time Backward** by one 19-year step and use the **Hand Tool** to measure the angular separation between the reference star and its reference position just above the NCP. Note this value in Data Table 1.
15. **Step Time Forward** by two 19-year steps to move the NCP to the opposite side of the reference star and repeat the angular separation measurement, entering this value in Data Table 1.
16. Translate these angles to arc seconds, average them, and divide the average by 19 to produce the angular rate of NCP motion in arc seconds per year.
17. Divide your result from the previous step by 3600 (the number of arc seconds in one degree) to obtain the angular rate of NCP motion, A, in degrees per year.
18. Follow the calculations outlined below to determine the revolutionary period of the precessional axis around a full circle.

DATA TABLE 1

Date	Angle between NCP and Star (',")	Angle (Arc seconds)	Average 19-year Rate (")	Rate Per Year ("/year)	Fraction of ° Per Year, A
1430 A.D.			D	D/19	D/(19 × 3600)
1468 A.D.					

Calculations

The measured angle between two successive 19-year positions of the reference star when the NCP is held stationary is equivalent to measuring the movement of the NCP against the star background. By measuring angular distance as a straight line we are assuming that this motion is straight, whereas the NCP is moving in a circle around the ecliptic pole, as we saw in section A. This movement can be assumed to be straight, however, because it is very small when measured over such a small time span.

We can now combine this motion with the equivalent radius of the precessional circle to determine the period of NCP rotation around this circle. The diagram below shows the geometry. O represents the center of the precessional circle, the north ecliptic pole. If the small base of the triangle PQ, of length A, is measured in the same units as the long side of the triangle, of length R, then the narrow angle θ subtended by this base at the apex of the triangle, in radians, is simply the ratio of the base over the side of the triangle. Because A actually represents motion of the NCP across the sky per year, θ will represent the angle of rotation of the NCP about the north ecliptic pole per year.

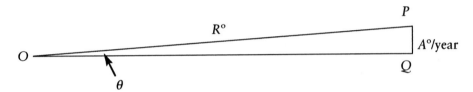

In the small-angle triangle *POQ*, the angle θ is given by

$$\theta \text{ (radians/year)} = \frac{A}{R}$$

1 radian = 57.3°. Thus, the angle through which the NCP moves around the precessional circle in degrees per year is

$$\theta \text{ (°/year)} = \frac{A \times 57.3}{R}$$

The inverse of this value, $\frac{1}{\theta}$, is the number of years taken to move through 1°. Consequently, to determine the time taken for the NCP to move through a full circle of 360°, we need to multiply $\frac{1}{\theta}$ by 360.

$$\text{Period for 1 revolution (years)} = (\frac{1}{\theta}) \times 360 \text{ years.}$$
$$= \frac{(R \times 360)}{(A \times 57.3)}$$

Because R = 23.45° is the angular distance between the NCP and the north ecliptic pole (i.e., the angle at which the rotation axis of the Earth is inclined to the perpendicular to the plane of the Earth's orbit around the Sun), the calculation becomes:

$$\text{Period for 1 revolution (years)} = \frac{(23.45 \times 360)}{A \times 57.3}$$
$$= \frac{147.33}{A}.$$

Question 8:   How fast does the North Celestial Pole move across our sky, in arc seconds per year?

Question 9:   How long will it take the spin axis of the Earth to move once around the precessional cone?

C. Motion of the Vernal Equinox

Another manifestation of the precessional motion of the sky is the year-by-year movement of the **vernal equinox,** one of the points of intersection of the **ecliptic** and the **celestial equator.** The vernal equinox defines the zero point of the right ascension-declination coordinate system. It is also important in the questionable practice of astrology, though practitioners of astrology do not follow the modern sky and its positions. Nevertheless, two references to the position of the Vernal Equinox—the first historic, the second modern—have achieved a certain significance. The vernal equinox is sometimes referred to by its historical name of the **First Point of Aries** because at a certain point in history, it was found to be within this constellation. We can determine the date when this label ceased to be relevant by tracing the era in which the equinox actually left the constellation of Aries. A more modern reference to the motion of the vernal equinox has entered the lexicon of modern English, namely the "Age of Aquarius," referring to the future. We can use *Starry Night*™ to predict the dawning of this age of infinite promise, which is presumed to start when the vernal equinox enters the constellation of Aquarius.

19. Select the view **Precession-C** under **Go/Observing Projects/Precession.**

You will note that, on the first day of spring in 2005 AD, the Sun is at the vernal equinox as expected. During the year, the Sun moves along the ecliptic at about 1° per year, but returns to the vernal equinox each year. However, precession will cause the position of the vernal equinox to drift slowly with respect to the background stars.

20. Start **Time Flow Forward** and observe the motion of the vernal equinox against the background stars. (You will note that the Sun and its accompanying planets move very rapidly across the sky during this rapid advance of time.)

You can now use this simulation to determine the two interesting dates referred to above, the date when the Vernal Equinox left Aries and the date when the Vernal Equinox enters Aquarius to begin the "Age of Aquarius."

21. Select the view Precession-C2 under **Go/Observing Projects/Precession.** To reduce confusion, the view does not show the Sun and planets.
22. Manipulate the time controls, time step interval, and zoom controls in the control panel to answer the following questions.

Question 10:   In what constellation is the vernal equinox located at the present time?

Question 11:   When did the vernal equinox, also called the First Point of Aries, actually leave the constellation of Aries?

Question 12:   When will we reach the "Age of Aquarius," defined by the vernal equinox moving into the constellation of Aquarius?

You will note that we are going to have to wait a significant time before we reach this age of great promise!

Because the vernal equinox is also the zero point of the celestial coordinate system, this motion changes the right ascension and declination coordinates of stars and other objects on a day-by-day basis. We can follow and measure this motion.

23. Select the view **Precession-C3** under **Go/Observing Projects/Precession**.
24. Position the cursor over the labeled star **HIP112813**, open the contextual menu and select **Center** to center the view on this star.
25. Measure the angular separation between this star and the vernal equinox.
26. Advance time by two steps of 19 years and repeat the measurement of the angular separation between the reference star and the vernal equinox.
27. Using the average of these two measurements, divide by 19 to obtain the rate of motion of the vernal equinox across our sky in arc seconds per year.

Question 13:   How fast does the vernal equinox travel across our sky in arc seconds per year?

D. Nutation

In addition to the slow but significant motion of the spin axis of Earth and the vernal equinox because of precession, the varying effects of the gravitational forces from the Moon and the Sun cause a much smaller wobble in the Earth's axis with a period reflecting the motions of the Moon. This wobble is known as **nutation** and can be observed and measured with *Starry Night*™.

28. Select the view **Precession-D** under **Go/Observing Projects/Precession.**

In this 1° field of view, a reference star, **TYC4662-45-1,** lies close to the NCP. In this mode, the reference system of NCP and meridian will remain fixed as time advances. Thus, although the NCP and meridian, which are tied to Earth, are in reality moving across the sky because of precession, they will appear to remain stationary while the stars will appear to move. Running Time Forward, you will see the precessional motion of the pole reflected in horizontal motion of the stars.

29. Start **Time Forward** and observe the motion of the stars.

You will note that, superimposed upon the horizontal precessional motion of the NCP shown by the horizontal motion of the stars, there is a small but perceptible wobble in the vertical positions of these stars as they traverse the sky.

30. Select the view **Precession-D** again under **Go/Observing Projects/Precession** and replay this sequence to show the effect of this **nutational** motion of the Earth's spin axis upon the effective motions of the stars in our chosen reference frame.

We can measure this nutational motion and quantify it. For this purpose, we will zoom in further on this region of sky to measure the motion of our reference star and thereby infer the wobbling of the Earth's spin axis.

31. Select the view **Precession-D2** under **Go/Observing Projects/Precession.**
32. Drag the screen so that the red line representing the celestial meridian is horizontal and the NCP is near the center of the view.

You can now make a series of measurements of the angular distance between the reference star, TYC4662-45-1, and the celestial meridian line. This line remains fixed in our sky and can be used as a reference line to measure the nutational wobble.

33. Use the **Hand Tool** to measure the perpendicular distance from the star to the celestial meridian. Note the date and your measurement in Data Table 2 below.
34. Advance **Time** in 2-year steps and repeat this angular measurement after each step.

DATA TABLE 2

Year											
Angle (")											

The data points in Data Table 2 can be plotted upon a sheet of graph paper to reveal the pattern of the nutation motion of the north celestial pole. Plot the angular separation in the vertical axis and the date along the horizontal axis. From the graph you can determine the relevant parameters of this motion: its period and its amplitude.

Question 14: What is the shape of the variation of the position of the celestial pole (i.e., is it a constant value, does it show linearly increasing variation, a sinusoidal variation, etc.)?

Question 15: If the star shows a periodic variation away from a straight-line motion, what is the period of this variation?

Question 16: What is the amplitude of this variation, if any? (Amplitude is one-half of the full variation of the angular separation of star from the reference line.)

E. Conclusions

The relatively smooth precessional motion of the Earth's spin axis, measured in Sections A, B, and C, is caused by the average effect of the differential gravitational forces from the Moon and the Sun acting upon the Earth's equatorial bulge. These forces do not always act in the same direction, however, because of the motion of the Moon along an orbital path that is inclined to the ecliptic plane. The points where the Moon's orbital path crosses the ecliptic, the nodes of the Moon's orbit, slide around the ecliptic plane with a period of about 18.6 years. As seen in Section D, this produces small differences in the gravitational influence of the Moon upon the Earth's bulge, resulting in the small nutational wobble superimposed upon the smooth precessional coning of the Earth's spin axis.

THE SUN'S POSITION AT MIDDAY

7

THE ANALEMMA

Midday is the point in the day at which the Sun is highest in the sky. This occurs when the Sun crosses an observer's celestial meridian. The celestial meridian, often called simply the meridian, is the imaginary line passing through an observer's zenith and the North and South Celestial Poles. Each point upon the Earth has a different zenith. Thus, your own meridian is specific to your position upon Earth. Your meridian passes vertically through the south point of your horizon and is the line across which objects pass at their highest elevation angle as they traverse your sky because of the Earth's rotation.

> The meridian is defined in Section 2–7 and illustrated in Figure 2–19 of Freedman and Kaufmann, *Universe*, 7th Ed.

Nevertheless, if we observe the Sun's position at 12 o'clock each day throughout the year, we will rarely find it on the meridian. There are several reasons for the Sun being offset from the meridian at noon.

The first reason is that the observer's location may not be at the center of the time zone. This produces a constant offset of the Sun from the observer's celestial meridian at noon. If time zone boundaries were ideal, this offset might be as large as half an hour. For political reasons, time zone boundaries sometimes differ from the ideal positions, resulting in offsets of close to an hour in some locations.

> Freedman and Kaufmann, *Universe*, 7th Ed., Figure 2–21 shows a map of Earth's Time Zones.

The second reason is that the Earth's orbit is elliptical, which means that the Earth's orbital speed and its distance from the Sun vary through the year, as described by Kepler's second law. The varying speed and solar distance cause the apparent speed of the Sun past the background stars to change day by day.

> Section 4–4 in Freedman and Kaufmann, *Universe*, 7th Ed., discusses Kepler's second law.

The third reason is that the Earth's spin axis is tilted with respect to the perpendicular to the plane of its orbit. This plane is known as the ecliptic plane. The tilt also causes a variation in the effective speed of motion of the Sun in a direction parallel to the Earth's equator. It is this tilt of the Earth's spin axis that also causes the variations of the seasons.

The combination of the Earth's axial tilt and its elliptical orbit causes a variable offset of the Sun from the meridian at midday (in addition to the constant offset due to position in the time zone), with the Sun drifting east and west in a cyclic pattern relative to the meridian over the course of a year. During this same time, the axial tilt also causes the Sun to drift north and south of the equator as the seasons progress. The result is a figure-8 pattern for the Sun's midday position in our sky over the course of a year. This figure-8 pattern is called the analemma, and is often found on globes representing the Earth.

> Section 2–7 in Freedman and Kaufmann, *Universe*, 7th Ed., discusses the effects of this deviation upon timekeeping.

The shape of the analemma was important when sundials were used to tell the time. The analemma provided corrections to sundial times amounting to as much as 16 minutes at certain times of the year. This observing project demonstrates the analemma by tracing the position of the Sun in the sky at noon for a whole year at a chosen site.

A. The Shape and Size of the Analemma

The first simulation will view the analemma from Calgary, Alberta, Canada, a city that lies at a longitude of about 114° W, within the Mountain Standard Time zone. The standard meridian for this zone is 105° west of the Greenwich Meridian. The Greenwich Meridian, also known as the prime meridian, passes through London, England and defines the zero point of longitude. Universal Time, a common standard time used throughout the world, is maintained at the Greenwich Meridian, 0 hours UT being midnight on this meridian.

Because the Earth rotates through 360° in 24 hours, or 15° per hour, the standard time in the Mountain Standard Time zone is 105/15 = 7 hours behind Universal Time.

In fact, Calgary is 9° west of the standard meridian for the Mountain Standard Time zone. Therefore, to place the Sun on the meridian in Calgary on average throughout the year we need to adjust the time to allow for this 9° offset. At 15° per hour, this offset amounts to 36 minutes of time.

1. Launch *Starry Night*™ and select **Analemma-A** under **Go/Observing Projects/Analemma**.

The view shows the southern horizon from Calgary on 12/21/2006 A.D., the day of the winter solstice. The local meridian is displayed as a line from the S point on the horizon to the zenith above the top of the screen. Time is set to 12:35:56 MST to place the Sun close to the meridian for Calgary's specific position within the Mountain Standard Time zone.

At this time in the year in the northern hemisphere, the Sun is at a low angle above the southern horizon. A vertical pole acting as a sundial at this location would cast a long shadow away from the S point on the horizon, toward the north.

2. To observe the change in position of the midday Sun at 3-day intervals through the year, click **Time Forward** and allow time to run for a year or two.
3. Stop time flow.
4. Select **View/Hide Daylight** and start time again to show the motion of the Sun against the background stars. Because the view direction is locked on to the meridian, the background stars appear to move as a consequence of the difference between the lengths of solar and sidereal days.

You can see that the position of the midday sun traces out a figure-8 pattern in a full year, climbing high in the sky in summer and returning to low elevation angles in winter. For a sundial to agree with our clocks, the Sun would have to be on the meridian at midday every day. It is obvious that this is rarely the case. The Sun is often quite far from this ideal position, leading to errors in timing from the use of a simple sundial.

You can make measurements of the analemma that allow you to answer several questions about the Sun-Earth system. For example, you can determine the tilt angle of the Earth's spin axis. You can also determine the errors in timing of an uncorrected sundial because of the irregularities of the Sun's apparent motion in our sky.

B. The Tilt of the Earth's Spin Axis

The reason for the north-south motion of the Sun is the tilt of the Earth's axis to the ecliptic plane. It is this tilt that produces seasonal changes upon the Earth. The maximum N–S excursion of the Sun in degrees of declination will be twice the tilt-angle of the Earth's spin axis to its orbital plane. We can measure this N-S excursion using the analemma figure.

5. Change the time step to **1 day** and use the time controls to move the Sun's position to the bottom of the analemma.
6. Open the **View Options** pane and clear the checkbox next to **Stars** under the Stars layer. This will hide the stars in the view. Then use the Angular Separation tool to measure the angle between the Sun and the top of the analemma. Note this angle in the data table below.

| Length of analemma (°/'/") | |
| Tilt of Earth's axis (°/'/") (analemma length × 0.5) | |

Question 1: What is the tilt angle of the Earth's spin axis from perpendicular relative to its orbital plane?

C. Errors in Sundial Times

The analemma represents the Sun's midday position with respect to an observer's meridian throughout the year. If you needed to use a simple sundial, such as the shadow of a vertical pole to tell the time (on a desert island as part of a television "reality" survivor program, maybe!), you would be in error on any given day by the time taken for the Sun to move through the angle represented by the difference between the analemma and the meridian. As you can see, your "clock" would be slow or fast, depending upon whether the point on the analemma was east or west of the meridian, respectively. We can measure these errors, using the traced analemma pattern.

7. Run **Time Forward** until the Sun is at the widest point of the analemma on the east (left) side of the meridian.
8. Measure the angle between the Sun and the meridian, note this angle in Data Table 1 below and record the date upon which this error was greatest.
9. Repeat this measurement for the west side of the analemma and record the error angle and the date.

The Earth rotates through 360° in 24 hours, the meridian moving across the background sky at a rate of 15° an hour. Thus, it takes 4 minutes of time for the meridian of the Earth to cover 1° of angle. The time error of a sundial in minutes of time can then be calculated easily by multiplying the angle in degrees and fractions of a degree, (the fraction being determined by dividing the number of minutes of angle by 60) by 4.

DATA TABLE 1

Date	Angle Error, ϕ	Sundial Error (minutes of time) {ϕ (°) × 4 minutes}

Question 2: Calculate the worst errors, in minutes of time, of the Sundial for the two dates of maximum error.

Question 3: At what times of the year will the sundial be most inaccurate? (In practice, determination of the date of maximum error is not very precise).

There are several days in the year when a sundial is accurate compared to our clocks.

10. Run **Time Forward**; stop the Sun at positions when the sundial error is zero and note these dates.

Question 4: On what dates in the year will a sundial be most accurate?

D. The Analemma from Other Latitudes

The shape of the analemma will be the same from any position on Earth. You can test this hypothesis by moving to another location.

11. With time flow stopped, change the Date to **June 21**. The specific year is not important.
12. Change the observing location to **São Paolo, Brazil**.
13. Turn off Daylight Saving Time if it is on and change the Time in the control panel to **12:00:00 P.M.**
14. Move the **Gaze Direction** to due **North**.
15. Drag the sky down so that the Sun is at the bottom of the view and the celestial meridian is vertical.
16. Run **Time Forward** to trace out the analemma from Brazil, and stop time flow when the Sun has traced out the full pattern.

TIP: Click an elevation button in the control panel
to erase previous analemma traces..

You will note that the shape of the analemma is the same as it was from your home location. Measurements equivalent to those carried out above will confirm that it has exactly the same dimensions when observed from this location. There is, however, one difference.

Question 5: What is different about the appearance and/or location of the analemma in the sky when viewed from São Paulo, Brazil, compared to that seen in Calgary? Why is this?

E. Analemma Details

In this section we will explore in more detail why sundials are in error when compared to clocks. As the previous simulations have shown, the Sun sometimes runs slow and at other times fast compared to the time on clocks. This is because clocks are built to run at a steady rate, calibrated to what is called the mean solar day. The duration of a mean solar day is the time required for a hypothetical Sun, moving eastward throughout the year at a steady rate, to return to the same position in the sky. Sundials, on the other hand, measure the apparent solar day. The length of the apparent solar day is the time required for the actual Sun to return to the same position in the sky as seen from a particular location on Earth. Sundial errors, as demonstrated by the analemma, represent the degree to which the actual Sun lags behind or runs ahead of the hypothetical mean sun through the year. These errors are sometimes known as the Equation of Time.

The two effects that combine to produce this offset of the Sun from its mean position day by day, the ellipticity of the Earth's orbit and the tilt of the Earth's spin axis to its orbital plane, are described in the introduction to this chapter. We can demonstrate the effect upon the analemma of each of these orbital parameters in turn by introducing a fictitious planet Vulcan into *Starry Night*™, whose initial orbital properties mimic those of the Earth. We can then remove these effects individually and observe the resulting effect upon the analemma.

The first of these effects can be removed by reducing Vulcan's orbital ellipticity to zero, making the orbit circular. The resulting analemma will show ONLY the effect produced by the tilt of the Earth's spin axis.

The orbit can then be restored to an elliptical shape and the tilt of the planet's axis to its orbital plane removed by moving this orbital plane to correspond to its equatorial plane. The subsequent analemma shape will be the result simply of the ellipticity of the orbit.

Finally, this planet's orbit can be made circular while maintaining the spin axis perpendicular to its orbital plane. You can then demonstrate that the analemma effect is removed entirely, the Sun returning precisely to the same position in the sky at the same time every day on this "ideal" planet.

We must first set up conditions in the program to introduce a new planet so that we can keep its properties but remove it after the project is completed.

17. Exit the *Starry Night*™ program without saving changes to the analemma file.
18. Open the folder in which your *Starry Night*™ software is located. Find and open the **Sky Data** folder and look for a file named **Other SS Objects.ssd**. If you find this file, rename it with the name **Other SS Objects.old**.
19. From the **Sky Data** folder navigate to **Go/Observing Projects/Analemma** and locate the file **Other SS Objects.ssd**.
20. Copy this file to the clipboard. Navigate back to the **Sky Data** folder and paste the file from the clipboard into this folder. This will introduce Vulcan into the *Starry Night*™ database of solar system objects.
21. Restart *Starry Night*™.
22. Select the view named **Vulcan** under **Go/Observing Projects/Analemma**.

The view is from the surface of Vulcan, a fictitious planet with orbital properties that closely mimic those of the Earth. Vulcan is on the Ecliptic Plane at a position very close to the Earth. (In fact, as we allow time to advance, the Earth will occasionally intrude upon your field of view!) You are at a longitude of 0° on the equator of this planet, and hence at a latitude of 0°, facing south but looking upward to your zenith. You will note that the time is 12:00:00 Universal Time (UT) on the Winter solstice, December 21, 2006 A.D. The Sun at this time is low on the meridian, and your zenith is in the center of the screen. Because the orbital parameters of this planet are very close to those of the Earth, advancing time in 3-day steps will demonstrate an equivalent analemma to that produced upon Earth.

23. Start **Time Flow Forward** and allow time to run for a year or two to trace out the analemma.

You will see that this analemma is the same general shape as that seen from Earth, having asymmetrical N and S lobes and an axis that is tilted slightly with respect to the meridian.

You can now adjust the ellipticity of the orbit of this planet to make its orbit circular.

24. Open the **Find** pane, right-click on Vulcan in the planet list, and select **Edit Orbital Elements...** to open the Orbit Editor window.

If you click on the Orbital Elements tab, a view of the solar system is displayed in which Vulcan's orbit is shown. The planet's position is close to that of the Earth. Most of the values of the orbital parameters set on the slide bars are not relevant to the present discussion but are the values that allow Vulcan to simulate the Earth.

Clicking on the Other Settings tab will show Vulcan with its axis tilted to the ecliptic plane at the same angle as the tilt-angle of Earth, 23.4°. Note that the Rotation Rate is set at 1.002738 rotations per day to account for the difference between the lengths of the solar day and the sidereal day.

25. Click the **Orbital Elements** tab and adjust the **Eccentricity** (e) to 0. Minimize the Orbit Editor window.

This step has reduced Vulcan's orbit to a circle. The resulting analemma will be the result ONLY of the effect of the tilt of the planet's axis to its orbital plane, in this case, the Ecliptic Plane.

26. Start the flow of time to observe the analemma, as seen from Vulcan moving in a circular orbit.

Question 6: What effect does eliminating the eccentricity from Vulcan's orbit have on the appearance of the analemma?

You can see that this analemma is now symmetrical, with equal N and S lobes. If you click the mouse momentarily anywhere on the sky, a small circle will show the position of your zenith. The analemma is centered upon this position in your sky.

You can now adjust the orbital parameters of Vulcan to demonstrate the effect of ellipticity only, without the tilt of its spin axis to the perpendicular to its orbital plane. The easiest way to accomplish this is to move the orbital plane of the planet to correspond to its equator. This will tilt Vulcan's orbital plane by 23.4° and move it away from the ecliptic plane. You will also need to restore the eccentricity of Vulcan's orbit to 0.017.

27. Restore the Orbit Editor window.
28. Under the **Orbital Elements** tab, enter 0.017 into the box for **Eccentricity.**
29. In the drop box labeled **Ref Plane,** select **Earth Equatorial 2000.**

Vulcan now has an elliptical orbit again but its spin axis is perpendicular to its orbital plane because the latter has been brought into coincidence with its equatorial plane, which is parallel to that of the Earth.

30. Minimize the Orbit Editor window.
31. Start the flow of time again and note the shape of the analemma under these orbital conditions.

You will see that the Sun now moves E-W across the meridian in a straight line around the zenith, taking a full year to complete two cycles of this motion. Thus the Sun, when viewed from a planet with its spin axis perpendicular to its orbital plane but with a slight eccentricity in its elliptical orbit, will appear to move only slightly around a fixed position in the sky.

We are now ready to move Vulcan into a perfectly circular orbit while its axial tilt remains at zero.

32. Restore the Orbit Editor window.
33. Adjust the **Eccentricity** to 0 and minimize the **Orbit Editor** window again.
34. Start the flow of time once more.

With Vulcan now moving in this "perfect" orbit, the Sun appears stationary with respect to the meridian at midday every day. Of course, the stars still appear to move in the background day by day because of Vulcan's motion in its orbit around the Sun.

Question 7: How accurate would a sundial be on the planet Vulcan as it is currently configured?

F. Sunrise Time at the Winter Solstice

The shape of the analemma can help to answer the following question often asked about sunrise time: "If the Sun is at its farthest South on the shortest day of the year, at the time of solstice on December 21, why is this NOT the date of the latest sunrise?" In fact, the date of latest sunrise is close to the last day of the year! We will define sunrise as the moment when the upper limb or edge of the Sun just touches the horizon. Because this time will depend to some extent upon the local horizon, the official definition assumes that the observer is at sea level. In order to simulate this situation, the present setup has a line representing a flat horizon.

35. Select the view **Analemma-F** under **Go/Observing Projects/Analemma**.

You are now back at Calgary at 8:45 A.M. on the day of the winter solstice facing SE, with the Sun in front of the line representing the horizon. You can check this compass direction by opening the **Info** pane and expanding the **Position in Sky** layer to display the Sun's **Azimuth**.

36. Run **Time Forward** and **Backward** in 3-day steps to trace the shape of the analemma at the winter solstice.
37. Stop time and adjust the date to December 21, close to date of the winter solstice. The actual year is not important.

In contrast to the view of the analemma at midday, this pattern is now tilted at a distinct angle to the horizon and is not perpendicular to it. Furthermore, the lowest point of the analemma pattern is no longer the position of the Sun at the winter solstice. The Sun will rise at the latest time of the day on the date when it is at its lowest point on the analemma, closest to the horizon. You can use *Starry Night*™ to determine the time and date of this latest sunrise by timing sunrise at various dates after the winter solstice, recording them in the table below.

38. **Center the Sun** in the view.
39. Adjust the time on this date to sunrise, with the horizon coincident with the upper edge of the Sun and record this time in Data Table 2. Note that, with the Sun centered on the screen, the horizon will appear to fall away to uncover the Sun at sunrise (which is precisely what happens in reality as the Earth rotates!).
40. Advance the date by 2 days, to December 23; again adjust the time to find the precise time of sunrise, and note this time in the Data Table.
41. Repeat this sequence until January 4.

DATA TABLE 2

Date							
Time of sunrise							

Question 8: What is the date of the latest sunrise in the winter of 2006–2007 A.D.?

You can readily see that the date of latest sunrise is many days later than the date of the winter solstice. The motion of the Sun because of Earth's orbital parameters, as represented in the shape of the analemma, affects the sunrise time. You can demonstrate this by running the analemma on the date of the latest sunrise.

42. Adjust the Date to that of latest sunrise and adjust the Time to sunrise.
43. Run the Date backward and forward by a few days to trace out the lower region of the analemma at this time. Note that, on December 21, about half of the Sun is already above the horizon at the time of latest possible sunrise.

Question 9: About how many days after the winter solstice does the latest sunrise occur?

Question 10: By how many minutes and seconds is this sunrise later than that at the solstice?

Thus, you can see how the irregularities of the Earth's motion around the Sun can affect the times of sunrise and sunset, moving the latest sunrise and earliest sunset away from the date of the shortest day near to the winter solstice. It would be a useful exercise for you to use the technique described above to demonstrate this equivalent effect upon sunset times. This can be done easily with the Sun centered upon the viewing screen. Simply change the Date to that of the solstice, switch the time to P.M., and slowly retard the time until the horizon covers the Sun. Back off the date in 2-day steps and find the sunset time for each date until the time of sunset reaches its earliest, around December 11. If you now change the field of view to about 20° and run the analemma, you will see how its shape and tilt to the horizon result in the earliest sunset many days earlier than the date of winter solstice.

There will be similar but smaller time shifts of the earliest sunrise and latest sunset around the time of longest day in the summer because of the shape of the analemma.

Question 11: Why will the effects of the Earth's irregular orbital motion upon the times of sunrise and sunset near summer solstice be smaller upon these times near to winter solstice? (*Hint:* Look at the shape of the analemma when the Sun is near to summer solstice.)

G. Cleanup

In order that the artificial planet Vulcan does not always appear whenever *Starry Night*™ is run in the future, it is important to remove it from the program database.

44. Exit *Starry Night*™.
45. Open the folder containing the *Starry Night*™ application and find and open the **Sky Data** folder.
46. Locate the file **Other SS Objects.ssd** and delete it.
47. If you find a file in the Sky Data folder named **Other SS Objects.old**, then rename this file **Other SS Objects.ssd**. This will restore solar system objects that you may have previously added to the *Starry Night*™ database.

H. Conclusions

In this project, you have demonstrated the combined effects of the Earth's elliptical orbit and the tilt of the Earth's spin axis to the perpendicular to this orbit upon the position of the Sun at midday through the course of the year and have seen this effect displayed as the analemma on the sky. You have measured the possible errors in a simple sundial because of these effects. These errors, when plotted for the whole year, are known as the **Equation of Time.** This "equation" is often displayed on modern sundials. You have also used the analemma to demonstrate that the date of latest sunrise in the winter is not the date of the shortest day of the year at the winter solstice.

THE MOON'S MOTION AND PHASES

8

The Moon orbits the Earth in 27.3 days with respect to the background stars. This period is known as a **sidereal month**. With respect to the Sun, the Moon orbits the Earth in 29.5 days, a time period known as the **synodic month**. The synodic month is longer than the sidereal month because the Earth-Moon system is orbiting the Sun while the Moon orbits the Earth. As an analogy, think of a clock with the Earth at its center and the numerals on the clock face representing the background stars. In this analogy, the hour hand can be thought to point toward the Sun's position in the sky while the minute hand points toward the Moon. At 12 o'clock, the hour and minute hands point in the same direction with respect to the numerals. After 1 hour, the minute hand has returned to the 12 o'clock position, but the hour hand has advanced to the 1 o'clock position. The minute hand must then continue its motion for just more than 5 minutes before it catches up with the hour hand again. The time taken for the minute hand to return to the 12 o'clock position is analogous to the sidereal period while the time taken for the minute hand to catch up with the hour hand is analogous to the synodic period.

> This cycle of lunar phases is described in Section 3–1 and Figure 3–5 of Freedman and Kaufmann, *Universe*, 7th Ed.

During the synodic month the Moon shows various phases as viewed from Earth. This arises from the fact that the Moon is a spherical body illuminated by the Sun. One-half of the Moon's surface is in light and the other half in shadow at any time. The proportion of the illuminated hemisphere visible from Earth depends upon the position of the Moon with respect to the orientation of the Earth and Sun.

In the monthly cycle of phases, the Moon appears first as a very thin crescent in the western sky, quite close to the Sun. It then proceeds to show more of its illuminated side to Earth as it moves farther from the Sun in the sky. In this **waxing**, or growing, half of the cycle it will pass through **crescent**, **first quarter**, and **gibbous** phases before reaching **full Moon**. At this point, the Moon is opposite to the Sun in our sky. Its angular separation from the Sun, measured across the sky, then begins to decrease. In this **waning**, or diminishing, half of the cycle, it proceeds through **gibbous**, **third quarter**, and **crescent** phases to **new Moon** again.

In this project, you will observe the motion of the Moon, observe its phases, and relate the phases to the position of the Moon with respect to the Sun in the sky.

A. The Moon's Motion Across the Sky and its Angular Speed

1. Launch *Starry Night*™ and open the **Find** pane. Expand the menu for the Moon using the blue down arrow and make sure that Enlarge Moon Size is checked.
2. Under **File/Preferences/Cursor Tracking (HUD)** check the **Name** and **Object type** options in the **Show** list.
3. Select the view **Go/Observing Projects/Moon's Motion/Moon-A.**

The view shows the sky looking south from Chicago at 12:45:00 A.M. Central Daylight Time on July 28, 1999. The full Moon is visible in the center of the view.

> 4. Change the time step interval to 300× real time and observe the motion of the Moon and stars in the night sky.

With time flowing at 300 times its normal rate, the westward drift of the sky resulting from Earth's eastward rotation becomes obvious. Our view of the sky from the rotating platform of Earth shows the stars and the Moon moving together in the opposite, or westward, direction.

The Moon shows an additional motion compared to the stars. If you watch the Moon carefully as the sky turns westward, you will see that it moves slowly eastward with respect to the stars. You can explore this further by adjusting the time and watching stars that appear to be close to the Moon on this date.

> 5. Select **Go/Observing Projects/Moon's Motion/Moon-A2.**
> 6. Note the faint star Sigma Capricorni just to the left (east) of the Moon. Center upon this star and change the field of view to about 5°.
> 7. Select a time step of 300× and watch the Moon as it moves in front of the star field and eventually covers this star at about 2:48 A.M. on 7/28/1999, in a process known as an **occultation.**

The sky appears to rotate and the Moon's path seems to be curved from this viewpoint during this sequence. This is an illusion caused by the Earth's rotation.

Observing the Moon at successive intervals of 1 sidereal day uncovers this independent motion of the Moon. At each observation 1 sidereal day apart, the stars will have returned to the same position in the sky. Any independent motion of the Moon will show up as a change in its position against the background stars.

> 8. Select **Go/Observing Projects/Moon's Motion/Moon-A.**
> 9. Click the **Single Step Forward** button to advance time by 1 sidereal day.

TIP: You can also use the **U** key on the keyboard to advance time in single steps of the time step interval. Use **Shift+U** to move back in time in single steps of the time step interval.

In the interval of 1 sidereal day, the stars have returned to the same position in the sky but the Moon, having proceeded some distance along its orbit around the Earth, has moved against the background stars.

> **Question 1:**   In which direction in the sky does the Moon move against the background stars?

We can measure how far the Moon moves in our sky in 1 sidereal day.

10. Select **Go/Observing Projects/Moon's Motion/Moon-A.**
11. Use the **Hand Tool** to measure the angular separation between the Moon and the star Albaldah. Note this value along with the date in the Data Table below.
12. **Single Step Forward** to advance time by 1 sidereal day and repeat the measurement of the angular separation between the Moon and Albaldah. Note the date and angular separation in the Data Table.
13. Calculate the difference between these two angular separations and enter this in the Data Table. The difference between these measurements is the average angular speed of the Moon over the chosen sidereal day.

DATA TABLE

Date	Angular Separation	Difference in Angular Separation

Question 2: What is the angular speed of the Moon against the background stars, in degrees per sidereal day?

Question 3: Approximately how long would it take for the moon to return to the same position relative to the stars?

The Earth's **orbital** motion around the Sun produces an apparent motion of the Sun against the background stars when viewed from Earth. In the course of a year, the Sun traces a complete path across the sky called the **ecliptic**. The ecliptic represents the intersection of the plane of the Earth's orbit with the celestial sphere. In the next sequence, we will track the Moon's motion with respect to the ecliptic.

14. Select **Go/Observing Projects/Moon's Motion/Moon-A3.**

The view shows the Moon from Chicago 4 sidereal days earlier than the view in **Moon-A.** The Moon is to the right in this view, above the line showing the ecliptic.

15. **Single Step Forward** in time again using steps of 1 sidereal day and watch the Moon's motion with respect to the ecliptic.

The Moon's orbit is inclined at about 5° to the ecliptic plane. Thus, the Moon is seen to follow a path across the sky that is close to the ecliptic, but does not exactly follow it. The Moon's path crosses the ecliptic plane at two points during each orbit around the Earth. These points are called the **nodes** of the Moon's orbit. The node at which the Moon crosses the ecliptic from south to north is called the **ascending node**, while that at which the Moon crosses the ecliptic from north to south is called the **descending** node. The nodes are significant because it is only when the Sun and the Moon are near to a node that **solar** or **lunar eclipses** can occur. Indeed, that is how the **ecliptic** got its name.

Question 4: What kind of path does the Moon take across the sky (e.g., curved, straight, etc.)?

Question 5: On what date does the Moon cross the ecliptic during the observing period in this sequence?

Question 6: Is this the ascending or the descending node of the Moon's orbit?

B. The Waxing Phases of the Moon

To demonstrate the phases of the Moon, we must observe the sky at different times of the night on different days. The best time to observe the waxing phases is just after sunset when the Moon appears in the western sky. The phases around full Moon are best observed near midnight, while the waning phases can be most effectively seen in the dawn sky.

16. Select **Go/Observing Projects/Moon's Motion/Moon-B.**
17. Select **View/Show Daylight.**

TIP: You can use Ctrl+D to turn Daylight on and off.

In this view of the sky looking west-northwest from Chicago at 8:30:00 PM, Daylight Central Time, on July 13, 1999, the Sun has just set. The Moon, barely visible as a faint sliver disk in the twilight, is labeled in the bottom center of the view. The portion of the Moon not lit by the Sun is visible from Earth at this time because of **earthshine.** This faint glow is caused by sunlight reflected from the surface and clouds of the Earth that illuminates the Moon. This appearance of the thin crescent Moon was referred to in earlier times as "the old Moon in the new Moon's arms." Note that the Moon is in the same general direction in the sky as the Sun at this time around New Moon.

18. **Single Step Forward** in time by **two** steps to advance time to **8:30:00 P.M.** on **July 15, 1999.** Notice that the time step interval is **1 solar day** rather than 1 sidereal day. Thus, the local time remains the same, close to sunset.
19. Move the **Hand Tool** to identify the bright object just below the Moon.

Question 7: What object appears near the Moon on July 15, 1999?

Question 8: In which way do the "horns" of the waxing crescent moon point with respect to the direction of the Sun?

Question 9: In which direction in your sky do the "horns" of the waxing crescent moon point, as seen from the Earth?

If you were to continue to step forward in time, the Moon would eventually disappear off the left side of the screen. To watch the Moon go through its phases, you will need to lock the view onto the Moon so that it remains centered in the screen as you manipulate the time flow.

20. Center the view on to the Moon.
21. Select **View/Hide Daylight** to turn daylight off again.
22. Advance **Time** by 1 day.

TIP: To center the view on the Moon, position the Hand Tool over the Moon, click the right mouse button, then choose Center.

Now, the Moon remains stationary in the center of the main view window while the sky *and the horizon* shift to the right (west). By keeping the Moon centered in the field of view the direction of your gaze shifts eastward, night by night.

23. Step forward to **July 20, 1999,** and watch the phase of the Moon change from day to day.

As time progresses and the Moon moves away from the Sun in the sky, the proportion of the Moon's disk illuminated by the Sun as seen from the Earth increases (waxes). On July 13, the Moon was a very slender crescent close to the Sun. By July 20, the Moon's orbital motion has carried it some distance away from the Sun in the sky and it shows the **first,** or waxing, **quarter phase.**

You can measure the angular distance between the Sun and the Moon as seen from the Earth and thereby gain some insight into how the relative positions of the Sun, the Moon, and the Earth produce the various phases of the Moon.

24. Adjust the **Time** to 8:00:00 P.M. and drag the view to the left until the Moon and the Sun are visible in the 100° field of view.
25. Measure the angular separation between the Sun and the Moon when the Moon is at first quarter phase.

Question 10:   What is the angular separation between the Moon and the Sun as seen from Earth when the Moon is at first quarter phase?

Question 11:   As seen from Earth, which side of the Moon's disk is illuminated by the Sun when the Moon is at the first quarter phase?

26. Center the view on the Moon once more and reset the time to **8:30:00 P.M.**
27. **Single Step Forward** in 1-day intervals to **July 27, 1999,** and observe the successive waxing gibbous phases of the Moon.

By July 27 the Moon is almost full. Notice that the Moon is close to the ESE point on the horizon, or 180° away from the Sun's position at WNW. For the Moon to be fully illuminated as seen from the Earth, it must be in the opposite part of the sky from the Sun.

C. The Waning Phases of the Moon

28. **Single Step Forward** 1 day to 8:30:00 P.M. on July 28, 1999.

The Moon is only just rising at 8:30 P.M. on July 28, 1999, whereas the sun has just set. This opposition of Sun and Moon in the sky is a condition for Full Moon. Because of the eastward motion of the Moon in its orbit around the Earth, it rises later each day. Clearly, you will need to change the time at which you make your observations during the waning phase to make sure that the Moon is above the horizon.

29. Change the time in the control panel to 01:00:00 A.M. **Daylight Saving Time** on **July 28, 1999.**
30. Watch the waning gibbous phase of the Moon as you step forward in intervals of one day to 01:00:00 A.M. on **August 4, 1999.** Identify the bright object that is visible just above Moon on August 4 at 01:00:00 A.M.

31. **Single Step Forward** 1 day to 01:00:00 A.M. on **August 5, 1999.** Identify the bright object above the Moon.

Question 12: What side of the moon's disk is illuminated by the Sun as seen from the Earth when the moon is at third quarter?

Question 13: In which direction, relative to the Sun, does the Moon move in the waning phases?

Question 14: Without using the Hand Tool to measure it, what do you expect to be the angular separation of the Moon and Sun as seen from the Earth when the Moon is at third quarter?

Question 15: To which object does the Moon pass close on August 4?

Question 16: To which object does the Moon pass close on August 5?

32. To continue to follow the waning phases of the Moon, change the time in the control panel to **5:45:00 A.M.** on **August 5, 1999.**
33. Watch the phase of the Moon wane from third quarter to crescent by stepping forward to **5:45:00 A.M.** on **August 9, 1999.**

Question 17: In which way do the "horns" of the waning crescent Moon point with respect to the direction of the Sun?

Question 18: In which direction do the "horns" of the waning crescent Moon point as seen from the Earth?

Question 19: Compare your answers to questions 8 and 9 with your answers to questions 17 and 18.

34. Single step forward in time 1 solar day to **5:45:00 A.M.** on **August 10, 1999.**

By 5:45:00 A.M. on August 10, the Moon is in the eastern sky, where it is rising just before the Sun. Thus, since July 13, it has nearly completed the entire cycle of phases back to new Moon. In fact, it will reach an equivalent position with respect to the Sun, just past new Moon, on about August 12.

D. Positions of the Sun, Moon and Earth and the Moon's Phases

In this section of the project, you will observe the phases of the Moon from a unique point of view that will allow you to see how the relative positions of Earth, Moon and Sun lead to the Moon's cycle of phases as these objects move through their respective orbits.

35. Select **Go/Observing Projects/Moon's Motion/Moon-D.**

The view is of the Sun and Moon from a location above the North Pole of the Earth on September 17, 2001. The Moon, labeled but virtually invisible, is very close to the Sun in the sky. Because the Moon is between the Sun and the Earth, the side of the Moon facing Earth is also facing away from the Sun and, as a consequence, is dark. This is the **new phase** of the Moon. The Moon and the Sun are in the same direction in the sky, as seen from the Earth.

36. Drag the sky upward until the Earth comes into view on the bottom of the screen to understand the orientation of your viewing location with respect to the Earth, Moon, and Sun.

Question 20: Considering the fact that the distance from the Earth to the Sun is roughly 400 times the distance from the Earth to the Moon, what geometrical term best describes the relative position of the Sun, Earth, and Moon when the Moon is in the new phase (e.g., right triangle, straight line, etc.)?

37. Select **Go/Observing Projects/Moon's Motion/Moon-D** to center the Moon in the view once again.
38. Advance time in single steps of 1 day until the Moon reaches first quarter phase.

You will note that, as the Moon moves away from the Sun direction, a sliver of reflected sunlight begins to be visible on the western edge, or limb, of the Moon as seen from the Earth. Stepping forward day by day will reveal progressively more and more of the sunlit side of the Moon. This is the waxing phase of the lunar cycle. As you can see, between 3 and 4 days after new Moon, the Moon is in the first or waxing crescent phase and seven days after New Moon, first quarter phase is reached. The relative positions of Moon and Sun can be seen by using the Hand Tool to drag the screen toward the left until the Sun becomes visible at the right (west) edge of the main view window. Because the screen encompasses a field of view of 100°, the Sun and the Moon are about 90° apart in the sky as seen from the Earth.

Question 21: What geometrical term best describes the relative positions of the Sun, Moon, and Earth at first quarter phase of the moon?

39. If necessary, center the view on the **Moon** once again.
40. Continue to advance time in 1-day steps and observe the Moon through the waxing gibbous phase to the point where the entire surface of the Moon that faces the Earth is now lit by the Sun, on **October 2, 2001 (10/2/2001 A.D.).**

This is **full Moon**, where again the Sun and Moon lie roughly on a straight line but now with the Earth between them. In contrast to new Moon, the angle between the Sun and the Moon is now approximately 180°.

41. Continue to advance time in single steps of 1 day to observe the Moon through waning gibbous, last quarter, and waning crescent phases to New Moon again on **October 16, 2001 (10/16/2001 A.D.).**

In the next sequences, you will make observations of the Earth from the Moon over the same period as the previous sequences in this section of the project.

42. Select **Go/Observing Projects/Moon's Motion/Moon-D.**
43. Position the **Hand Tool** over the Moon, open the contextual menu for the Moon, and select **Go There.**
44. Open the **Find** pane and double-click the entry for the Earth to find and center the view on the **Earth.**

The time is the same as that of the beginning of the previous section, when the Moon as seen from Earth was at new phase. Now, the view is of the Earth as seen from the Moon.

45. Zoom in on the view of the Earth.

> **Question 22:** At this time, when the Moon as seen from Earth is in its new phase, at what phase is the Earth as seen from the Moon?

46. **Advance time** in single steps of 1 day to the date of full Moon as seen from the Earth, namely **10/2/2001,** and observe the changing phase of the Earth as seen from the Moon.

> **Question 23:** Is the Earth waxing or waning?
>
> **Question 24:** What is the phase of the Earth as seen from the Moon on the date at which the Moon is full as seen from Earth (10/2/2001)?
>
> **Question 25:** On 10/2/2001, in which direction in the sky is the Earth, as seen from the Moon, with respect to the Sun?

47. **Advance time** in single-day steps to the date of the next new Moon **(10/16/2001).**

> **Question 26:** In this period, while the Moon wanes as seen from Earth, what phases does the Earth go through as seen from the Moon?

E. The Synodic Month

48. Select **Go/Observing Projects/Moon's Motion/Moon-E.**

The view is from Chicago on July 12, 1999. The horizon and daylight have been hidden to give an unobstructed view of the Moon. The new Moon appears near to the Sun in the sky. Note the background of stars that surround the Moon, particularly the position of the bright star **Mekbuda,** which is labeled.

The time required for the Moon to complete a full cycle of phases from new Moon to new Moon is called the **synodic month.** Because the phases of the Moon depend on its relative position with respect to the Sun as seen from the Earth, the synodic month is also the time required for the Moon to make one complete orbit with respect to the Sun as viewed from the Earth.

49. Watch the complete cycle of lunar phases by stepping through time in 1-day intervals. Stop after 27 such daily steps and note the position of the Moon relative to the reference star Mekbuda. A little more than 27 days, or 1 sidereal month, was required to bring the Moon back to the same region of the sky with respect to the background stars.
50. Advance **Time Forward** by two more steps of one day to bring the Moon back to new Moon phase. You will see that about 29.5 days, or 1 synodic month, would have brought the Moon back to the same position with respect to the Sun.
51. Select **Go/Observing Projects/Moon's Motion/Moon-E** and repeat the sequence to answer the following questions

Question 27:   How many days elapse between:
                     a. New Moon and first quarter?
                     b. New Moon and full Moon?
                     c. New Moon and the next new Moon?

Question 28:   How long, in days, is a synodic month?

52. Select **Go/Observing Projects/Moon's Motion/Moon-E.**
53. Open the **View Options** pane and select **The Ecliptic** under the **Guides** layer.
54. Position the **Hand Tool** over the Moon and open the contextual menu. Select **Orbit** to show the orbit of the Moon in the view.
55. Make sure that the view is centered on the Moon.
56. Step **Time Forward** in 1-day steps to **8/10/1999** and observe the Moon's path relative to the ecliptic, and then answer the following questions.

Question 29:   How often does the Moon cross the ecliptic plane in one synodic month?

Question 30:   On what date does the Moon pass through the ascending node of its orbit?

Question 31:   On what date does the Moon pass through the descending node of its orbit?

E. Setting of the Sun and Crescent Moon

In addition to being very informative, watching the motion of the Moon in the sky can also be very beautiful. One of the finest sights that occur every month is the setting Sun flanked by a new crescent Moon cradling the rest of the Moon, visible by the ghostly illumination of earthshine, as they slowly slide below the horizon in a dark blue twilight.

57. Select **Go/Observing Projects/Moon's Motion/Moon-F.**
58. Change the time step interval to **300×** real time and observe the events in the western sky.

Question 32:   What is the time of sunset?

Question 33:   What is the time of moonset?

F. Lunar Occultation of a Star

One striking consequence of the lunar motion against the background stars is the occasional passage of the Moon in front of a star, an event known as an **occultation**. This phenomenon was briefly mentioned in Section A. This is an exciting event to watch through a telescope. As the Moon slides eastward against the background stars it occasionally covers and uncovers stars in its path. When the Moon covers a star, the event is called a disappearance and when it uncovers a star, this is known as a reappearance event. These events are particularly exciting to watch when they occur against the unlit limb (edge) of the Moon. Because the stars are (almost) true points of light in the sky, and because the Moon has no atmosphere, the star vanishes almost instantaneously from view in a disappearance event against the dark limb of the Moon. In fact, a star's angular size can sometimes be determined by the accurate measurement of the time taken for a star to disappear during an occultation. The following setup will simulate these occultation events for an almost full Moon.

59. Select **Go/Observing Projects/Moon's Motion/Moon-G.**
60. Examine the control panel. The initial conditions are almost the same as those in section A of this exercise except that the time is 5 minutes earlier, the field of view is locked on the Moon and encompasses only 3°, and the time step interval has been set to 1 second.
61. Change the time step to 300× and observe the motion of the Moon against the background stars. You will note that several faint stars become occulted by the dark advancing limb of the Moon.

An occultation event is particularly interesting when circumstances are such that only the mountains and crater walls of the Moon's southern or northern polar regions pass in front of the star. This is what is known as a **grazing occultation** and, from the correct location on Earth, the star is seen to wink out several times as it moves behind these higher regions at the Moon's poles. This type of occultation provides an opportunity for observers to make important observations of regions of the Moon that are otherwise difficult to measure. If several observers watch the occultation through telescopes from sites spaced at right angles to the shadow path of the moon across the Earth and carefully time the disappearances and reappearances of the star, a detailed profile can be constructed of the topography along the Moon's limb.

A similar procedure can be used when an asteroid occults a star. Asteroids are too far away for us to be able to see them as much more that just points of light, even through the largest telescopes. However, occultation observers scattered across the shadow path of the asteroid on the Earth see the different parts of the asteroid passing in front of the star. Careful timings of the disappearance and reappearance of the star from several different locations can tell us the shape and size of the asteroid. A star has occasionally been seen to disappear and reappear a second time for a particular asteroid, indicating that this asteroid is accompanied by a small moon.

> Question 34: An occultation event can be a disappearance or a reappearance and can occur against the bright limb or the dark limb of the Moon. For each combination below, identify whether the Moon is waxing or waning.
> a) Disappearance at the dark limb
> b) Reappearance at the bright limb
> c) Reappearance at the dark limb
> d) Disappearance at the bright limb

H. The Earth from the Moon

It is interesting to use *Starry Night*™ to observe the phases of the Earth as seen by astronauts on the Moon.

62. Select **Go/Observing Projects/Moon's Motion/Moon-H.**

Examine the control panel. Notice that the current viewing location is about 100 km south of the crater Herodotus on the Moon. The time-step interval is 1 hour.

63. Click **Flow Time Forward** to watch the horizon and the Earth as time progresses through at least one full cycle of phases of the Earth.

This provides a fascinating view of our Earth as it moves against the background sky as seen from the Moon. You can see it rotate on its axis and show phases equivalent to those of the Moon when seen from Earth.

An interesting question arises in this sequence, as time progresses in intervals of 1 hour and the view remains centered upon the Earth. You will note that the Earth's position does not change with respect to the lunar horizon.

Question 35: Notice that the horizon does not get in the way in this sequence. Can you explain why?

Question 36: Why does the Earth remain in the same position in the lunar sky for a long period from any given location?

I. Interesting Historical Set-up

On July 20, 1969, at 4:17:40 P.M. Eastern Daylight Time (20:17:40 UT), Neil Armstrong and Buzz Aldrin landed the lunar module *Eagle* on the surface of the Moon at Tranquility Base (lunar coordinates: 0°4'5" N, 23°42'28" E). At 10:56:15 P.M. Eastern Daylight Time (02:56:15 UT July 21, 1969), Neil Armstrong stepped off of the footpad of the lunar module onto the surface of the Moon.

Configure *Starry Night*™ so that you can see the Earth as Neil Armstrong would have seen it at the moment tht he said, *"That's one small step for [a] man—one giant leap for mankind."*

> **TIP:** *Starry Night*™ includes all of the Apollo landing sites in the location list for the Moon.

Question 37: In what phase did Neil Armstrong see the Earth? (*Hint:* Select **Go/Observing Projects/Moon's Motion/Moon-I**, center the view on the Earth and zoom in.)

Question 38: What was the phase of the Moon from Earth at that moment? (*Hint:* You can select **Go There** from the contextual menu for the Earth to confirm your answer.)

J. Conclusions

In this project, you have observed the phases of the Moon from different viewpoints as the Moon moves around the Earth in its orbit, while noting the relative positions of the Moon and the Earth with respect to the Sun in the sky for each phase. You have measured the Moon's angular speed across our sky and you have compared the sidereal and synodic periods of the Moon as measured from the Earth. You have also examined the phases of the Earth as they would appear to an explorer on the Moon and have seen that these phases are complementary to those of the Moon as seen from Earth.

SOLAR ECLIPSES

9

It is a fortunate coincidence that the angular sizes of the Sun and the Moon are about equal in our sky. Consequently, when they are aligned, the Moon can obscure (or occult) the Sun and produce a **solar eclipse**. As you recall from the project on the Moon's Motion, it is only when the Moon is new that it is in the same direction in the sky as the Sun as seen from Earth. If the plane of the Moon's orbit were exactly in the Earth's orbital plane, the ecliptic plane, then a solar eclipse would occur at every new moon. In fact, the orbital plane of the Moon is inclined to the ecliptic plane by about 5°. Thus, only when the Sun is close to the points in the sky where the Moon is crossing the ecliptic plane will it occult some portion of the Sun. These positions are known as **nodes** of the Moon's orbit.

> Section 3–3 of Freedman and Kaufmann, *Universe*, 7th Ed., discusses eclipses in general, while Section 3–5 discusses solar eclipses in particular.

The Sun is close to a node twice per year. An eclipse can occur during a period known as an **eclipse season** that lasts for about a month around this time. Because of the precession of the Moon's orbit, eclipse seasons are a little less than 6 months apart. They move slowly through the calendar, occurring earlier each year.

The geometry of Earth, Moon, and Sun is such that the length of the Moon's shadow is just less than the average distance between Earth and Moon. Consequently, even though the Moon and Sun are aligned in the sky, the Moon's shadow cone does not always reach the Earth. Under these conditions, observers in the path of the Moon's shadow will see the Moon somewhat smaller than the Sun and, at the moment of maximum eclipse, a ring of the Sun's disk will still be seen. This is known as an **annular solar eclipse**. Observers who are outside the central region of the eclipse will see only part of the Sun's disk covered and will therefore be in the **partial eclipse** zone.

Because the Moon's orbit is elliptical, there are times when the Moon is closer than average to the Earth. If this occurs when the Moon is new and the Sun is at or near a node, the darkest part of the Moon's shadow cone, the **umbra**, reaches the Earth. Observers within the narrow shadow region will see a **total solar eclipse**; the Moon completely obscures the Sun for a period that can last more than 7 minutes. Just as in an annular eclipse, observers on either side of the path of totality will see a **partial solar eclipse** where the Moon covers only part of the Sun's disk.

The total eclipse of the Sun is one of nature's finest spectacles. Only at this time can you see the tenuous but hot outer atmosphere of the Sun directly without protection for the eyes or special instrumentation. Indeed, this geometry is such that many phenomena can be seen and measured ONLY during this brief eclipse period known as **totality**. The Sun's outer atmosphere directly above the Sun's visible surface is composed of the **chromosphere** overlaid by a very tenuous but extremely hot outer region called the **corona**. Both have been studied extensively over the past century using eclipse observations.

A. Apparent Sizes of the Sun and Moon

The radius of the Sun is nearly 696,000 km, while that of the Moon is 1738 kilometers. On the other hand, the Sun is at an average distance of about 150 million km from the Earth, whereas the average distance from the Earth to the Moon is about 384,500 km.

> **Question 1:**   What is the ratio of the radii of the Sun and the Moon?
>
> **Question 2:**   What is the ratio of the average distances of the Earth from the Sun and the Moon?
>
> **Question 3:**   How do the two ratios compare?

While the Sun is very much larger than the Moon, its greater distance from Earth reduces its apparent size to match that of the much smaller Moon. The result is that the Sun and the Moon appear to be roughly the same size in our sky. However, the orbit of the Moon around the Earth and the orbit of the Earth around the Sun are elliptical in shape. When the Moon is closest to the Earth, at perigee, it will appear larger in the sky than when it is farther from the Earth, at apogee. The line across the Moon's orbit connecting apogee and perigee is called the **line of apsides.**

1. Launch *Starry Night™*.
2. Select the view **Eclipse-A** under **Go/Observing Projects/Solar Eclipses.**

The view is centered on a close-up of the nearly full Moon. Note the proportion of the 1° field of view that the Moon occupies.

3. Start **Time Flow Forward** and watch the apparent size of the Moon change over time. In this simulation, note that the contrast between the dark and bright parts of the Moon as it progresses through its phases has been artificially decreased.

> This dark side brightness can be adjusted if necessary by opening the **View Options** pane, positioning the cursor over **Planets–Moons** in the **Solar System** layer, and clicking the **Planets–Moons Options** button that appears. A dialog window opens that includes a slide bar control for adjusting the dark side brightness.

In the previous step, you should have noticed that the overall apparent angular diameter of the Moon appears to grow and shrink in cycles over time. As the Moon orbits Earth in an elliptical orbit, it appears larger at perigee and smaller at apogee.

4. Select the Moon, open the **Info** pane for the Moon, and expand the **Position in Space** layer and the **Other Data** layer.

> **TIP:** To select the Moon, position the Hand Tool over the Moon, open the contextual menu and choose Select Moon.

5. Select **Eclipse-A** under **Go/Observing Projects/Solar Eclipses**.
6. Single-step forward in time in intervals of 1 day while noting the **Distance from Observer** in the Info pane. Find the date at which the Moon's distance is the greatest. Record the Date, Angular Size of the Moon as seen from the Earth, and its Distance from Observer in Data Table 1.
7. Single-step forward 1 day at a time and find the date at which the Distance from Observer is least. Record the Date, Angular Size, and Distance in Data Table 1.
8. Single-step forward 1 day at a time to find the next date at which the Moon is again farthest away. Record the Date, Angular Size, and Distance in Data Table 1.

DATA TABLE 1

Date	Angular Size of Moon (')	Distance from Observer (km)

Question 4: What is the approximate range of the apparent angular size of the Moon in the sky?

Question 5: Approximately how many days does it take for the Moon to go through a complete cycle from smallest apparent size through largest apparent size and back to its smallest apparent size?

Question 6: What is the relationship between the Moon's angular size and its distance from Earth?

Just as the Moon appears alternately larger and smaller through the month due to the elliptical orbit of the Moon around the Earth, so the Sun appears larger and smaller in our sky as the Earth moves through its elliptical orbit around the Sun. When the Earth is closest to the Sun, at perihelion, the Sun will appear slightly larger in the sky than when the Earth is at aphelion, the farthest point in its orbit from the Sun.

9. Select **Eclipse-A2** under **Go/Observing Projects/Solar Eclipses**.
10. Start **Time Flow Forward** and observe the size of the Sun in the view.
11. Select the Sun in the view, open the **Info** pane and expand the **Position in Space** and **Other Data** layers.
12. Select **Eclipse-A2** under **Go/Observing Projects/Solar Eclipses**.
13. Verify that the Sun is closest to the Sun on this date by moving forward and back in time in 10-day intervals.
14. Note the Date, and the Sun's Angular Size and Distance from Observer in Data Table 2.
15. Single-step forward in time in intervals of 10 days and find the date at which the Sun is farthest from the Earth. Record the Date, Angular Size, and Distance from Observer in Data Table 2.

DATA TABLE 2

Date	Angular Size of Sun (')	Distance from Observer (au)

Question 7:   In what month does the Sun appear smallest in our sky?

Question 8:   In what month is the Earth at aphelion?

Question 9:   In what month does the Sun appear largest in our sky?

Question 10:   In what month is the Earth at perihelion?

Question 11:   What is the approximate range in angular size of the Sun through the year?

B. Eclipse Geometry

> Figure 3-7 of Freedman and Kaufmann, *Universe*, 7th Ed., shows the conditions required for an eclipse to occur and also illustrates the Line of Nodes.

In a solar eclipse, the Moon passes in front of the Sun, occulting part or all of the Sun's disk as seen from a particular location on the Earth. As you recall from the project on the Moon's Motion and Phases, when the Moon is in the same general direction in the sky as the Sun, it is at its new phase. However, because the Moon's orbit is inclined to the ecliptic by about 5°, the Moon usually passes north or south of the Sun when it is at new phase and so fails to produce an eclipse. A solar eclipse can occur only when the Moon crosses the ecliptic at one of the nodes of its orbit. Because the Sun and Moon must be aligned for an eclipse to occur, the Sun must also be in the direction of one of the nodes of the Moon's orbit.

16. Select the view **Eclipse-B** under **Go/Observing Projects/Solar Eclipses**.

The view shows the Sun centered on the screen, flanked by Mercury on the right and Venus on the left. The green line arching over the Sun shows the path of the Moon's orbit around the Earth. The small vertical bar on this line indicates the perigee point of the Moon's orbit. Toward the left, on the line of the Moon's orbit, you will notice a hollow wedge shape showing the point of the **descending node** of the Moon's orbit.

17. Change the Date in the control panel to **December 16, 1991 (12/16/1991 A.D.)**.

Once again, a small wedge icon appears to the left of the Sun on the line representing the Moon's orbit. However, now the wedge is solid. The solid wedge represents the **ascending node** of the Moon's orbit.

18. Use the time controls in the control panel to position the Sun in the same direction as one of the nodes of the Moon's orbit. Record the date of this event in Data Table 3.
19. Run **Time Forward** to the date at which the Sun crosses the next node. Record the date in Data Table 3. Repeat this step several more times.
20. Determine the difference between these successive dates to find the time intervals between node crossings.

DATA TABLE 3

Date of Node Crossing	Intervals Between Node Crossing (days)

Question 12:   On average, how many days elapse between the dates when the Sun is at a node of the Moon's orbit?

During an eclipse season, when the Sun is close to one of the nodes of the Moon's orbit, some type of solar eclipse is inevitable. There are two such seasons in an eclipse year, one when the Sun is near the ascending node and one when the Sun is near the descending node.

Because the Moon and the Sun are extended objects in the sky, they do not have to meet precisely at a node for an eclipse to occur. An eclipse season begins when the Sun is about 15° from a node and concludes when the sun is about 15° past a node along the ecliptic. Whenever the Sun is within this range, it is possible that a new moon will occult some portion of the Sun's disk as seen from a particular location on the Earth.

Question 13:   What is the approximate duration, in days, of an eclipse season? In other words, for how many days is the Sun within 15° of a node?

Question 14:   What is the synodic period of the Moon?

Question 15:   Compare your answers for the previous two questions. What is the minimum number of solar eclipses that can occur in an eclipse season?

Question 16:   What is the maximum possible number of solar eclipses that can occur in an eclipse season?

Question 17:   What is the minimum number of solar eclipses that must occur in a calendar year?

Question 18:   What is the maximum possible number of solar eclipses that can occur in a calendar year?

C. Precession of the Lines of Nodes and Apsides

If the Moon's orbit were fixed, then an eclipse year would be the same as a calendar year. However, gravitational effects from the Sun cause the Moon's orbit to precess. As a result of this precession, the nodes drift slowly westward over time, the effect being known as the regression of the nodes. Consequently, eclipse seasons are somewhat less than 6 months apart and an "eclipse year" is shorter than a calendar year. The next sequence demonstrates this westward precession of the nodes of the Moon's orbit.

21. Select the view **Eclipse-C** under **Go/Observing Projects/Solar Eclipses.**

The view is centered on the Earth and shows the orbit of the Moon from a position about 0.01 AU above the Earth and nearly perpendicular to the plane of the Moon's orbit.

Recall that the "wedge" symbols on the display of the Moon's orbit represent the nodes of the Moon's orbit. The line of nodes is an imaginary line connecting these nodes. The small bar perpendicular to the line showing the Moon's orbit represents the position when the Moon is at perigee and a line from this position through the Earth to the apogee position is known as the **line of apsides.**

22. Run **Time Forward** and carefully observe the changing position of the line of apsides relative to the stars and to the line of nodes.
23. Stop time when one of the nodes is near a star. Adjusting the time step as desired (20 sidereal days is a good choice) to estimate the time required for the line of nodes to complete a full revolution. Use a similar procedure to time the period of the line of apsides.

Observe the way the orientations of the line of nodes and line of apsides vary over time. Note that the choice of Time Step in intervals of sidereal days serves to maintain the star background stationary while we examine the motions of lines of nodes and apsides.

The line of nodes is important in defining the times of the eclipse seasons when eclipses are possible. We see that the position of the nodes moves westward with time. Thus, the Sun, as it moves eastward around the ecliptic, will reach these positions somewhat earlier than if they were stationary, making the eclipse year somewhat less than a calendar year.

The line of apsides defines the positions when the Moon will be at its nearest and farthest from Earth and this will define when the Moon will have its largest and smallest angular diameters in our sky. This will in turn determine which eclipses are total and which are annular.

Question19 :   Approximately how long does it take for the line of nodes to complete a full cycle with respect to the star background?

Question 20:   How long does it take for the line of apsides to complete a full cycle with respect to the star background?

D. The Shadow of the Moon

See Figure 3–11 of Freedman and Kaufmann, *Universe*, 7th Ed., for an illustration of the shadow cone of the Moon.

Although the Moon and the Sun appear to be the same size in our sky, the Sun, being much more distant, is proportionally larger. Consequently, the Moon's shadow is in the shape of a cone. In a solar eclipse, the Earth passes into some part of the Moon's shadow cone and all or a part of the Moon's circular shadow is cast upon the surface of the Earth.

In the next sequence, you will observe a partial solar eclipse from a viewing location some distance from the center of the shadow. In the simulation of a partial eclipse that *Starry Night*™ produces, you will be able to see the Moon as it approaches the Sun. In the real sky, the Moon is not visible until it actually encroaches upon the Sun's disk.

IMPORTANT! Never look directly at the real Sun, even when it is partially eclipsed, without proper eye protection (Welder's glass #14 or specially designed aluminized glass or plastic filters). In the absence of proper eye protection, observe a projected image of the Sun. **Looking at the Sun without proper protection can produce serious and permanent eye damage!**

24. Select **View from Cairo** under **Go/Observing Projects/Solar Eclipses**.

The view is from Cairo, Egypt, and is centered on the Sun at 11:00 A.M. on October 3, 2005. The Moon is visible in the view, above and to the right of the Sun.

25. Select a time step of 300× real time and observe the partial solar eclipse.

In the next sequence, you will observe this same event from a position in space about 0.001 AU over Cairo.

26. Select the view named **Cairo from Space** under **Go/Observing Projects/Solar Eclipses**.

The view is looking down on Earth from a position directly over Cairo, Egypt. In the upper left quadrant of the Earth, the outline of the Moon's penumbral shadow is illustrated.

27. Select a time step of 300× real time and observe the Moon's shadow track across Africa.

The area inside the expanding line defines the region upon Earth where at least part of the Sun is eclipsed while the dark patch in the center of this area is where the central region of the Sun is eclipsed, along the path of totality.

As you learned in the previous section, the Sun needs to be within only about 15° of one of the nodes of the Moon's orbit for a solar eclipse to occur. Whenever the Sun is within about 10° of a node at the time of new moon, the alignment of the Moon and Sun as seen from Earth is close enough that the central portion of the moon's circular shadow touches some point on the Earth.

The central part of the shadow cone is the darkest and is known as the **umbra**. From anywhere within the umbra, the Moon would be seen to completely blot out the sun. Surrounding the umbra is the penumbra, where light from some portion of the Sun is still visible, producing a solar eclipse that is only partial (i.e., the Moon covers only part of the Sun).

The average length of the Moon's umbral shadow is somewhat shorter than the average distance from the Earth to the Moon. Even in the center of the shadow, the factors governing the apparent size of the Sun and Moon as seen from Earth can conspire to produce a central eclipse that is not total. In other words, although the center of the Moon passes very nearly across the center of the Sun as seen from some location on Earth, the Moon is far enough from the Earth to appear smaller than the Sun. At maximum eclipse, a thin, bright ring of the Sun's disk surrounds the occulting Moon. As described in the introduction, this type of central eclipse is called an **annular eclipse**.

28. Select the view **22N, 21E** under **Go/Observing Projects/Solar Eclipses**. Examine the control panel.

Figure 3–12 of Freedman and Kaufmann, *Universe*, 7th Ed., shows a photograph of an annular solar eclipse.

You are now viewing the same eclipse that you observed as a partial solar eclipse from Cairo, Egypt, in a previous sequence but from a slightly different location.

29. Select a time step of 300× and set **Time** running forward (or advance the time by minutes) to observe the eclipse from this location.
30. Stop the Time advance at maximum eclipse and note down this time.
31. Select the view **22N, 21E from Space** under **Go/Observing Projects/Solar Eclipses**.
32. Select a time step of 300× and observe the track of the Moon's shadow as seen from a point in space directly over the intersection of Latitude 22° N and Longitude 21° E (1h 24m E).
33. Select the view **22N, 21E from Space** under **Go/Observing Projects/Solar Eclipses**. Open the Planets options dialog with **View Options/Solar System/Planets-Moons**. Check the **Surface Guides** option and all of its suboptions. Click **OK**.
34. Select a time step of 300×.
35. Using the surface grid as a guide, mark the location upon Earth at Latitude 22° N, Longitude 21° E (1h 24m E) with the **Hand Tool** cursor. Observe the track of the Moon's shadow relative to this location upon Earth.

36. Adjust the time to that measured for maximum eclipse at this location and note that the center of the Moon's shadow lies directly over the chosen location upon the Earth.

37. Select the view **Cairo from Space from Go/Observing Projects/Solar Eclipses.** Turn on the **Surface Guides** option in the Planet options dialog as in step 33.

38. Select a time flow of **300×** and observe the track of the Moon's shadow relative to Cairo at Latitude 30° N, Longitude 31° E (2h 4m E).

E. Total Solar Eclipses

> Figure 3-10 of Freedman and Kaufmann, *Universe*, 7th Ed., is a photograph of a total solar eclipse.

Occasionally, the apparent size of the Moon and the Sun and their alignment from certain locations on Earth are such that the Moon completely blots out the bright disk of the Sun at the midpoint of a central eclipse. As the Moon orbits eastward around the Earth, the darkest part of the Moon's shadow, the umbra, traces a path across the surface of the Earth. Observers on the Earth who are located along the centerline of the shadow will witness a total solar eclipse.

A total solar eclipse is one of the most awesome sights of nature. Many people travel around the world chasing these exotic events.

There are four distinct contact points in a total eclipse. First contact occurs when the Moon first encroaches on the Sun as the edge of the Moon's penumbral shadow just begins to cross the viewing location. After first contact, the Moon gradually covers more and more of the Sun's disk during the partial phase of the eclipse. As seen from space, the viewing location on the Earth becomes more deeply engulfed in the penumbral shadow of the Moon. Only when the Moon has covered about 90% of the Sun's disk will observers on the ground notice a significant diminution in the brightness of the daylight. The sky becomes a very dark twilight blue. When only the smallest sliver of the sun's disk remains uncovered, the topography of the Moon becomes evident as irregularities in the crescent of remaining sunlight. As the Moon continues toward a complete occultation of the Sun, a few last rays of light from the Sun may pass through lunar valleys and strike the Earth. The result, as seen from the Earth, is a string of bright beads of sunlight along the advancing limb of the Moon. These are called Baily's Beads. When only one such bead remains, the shadow has deepened sufficiently that the entire disk of the Moon appears encircled by the silvery glow of the solar corona while the bright solitary bead of photosphere sparkles like a diamond. The effect is called, aptly, the diamond ring effect.

At second contact, the Moon completely covers the Sun's disk and the marvel of totality unfolds. Only at this time is it safe to look directly at the Sun without eye protection, and one is encouraged to do so. With the bright photosphere of the Sun completely blocked out by the Moon, the relatively fainter atmosphere of the Sun (the chromosphere and corona) becomes visible. As seen from space, locations on the Earth's surface that lie in the path of totality are within the Moon's umbral shadow.

At third contact, totality ends and it is once again important to use proper eye protection while observing the final partial phase of the eclipse. The diamond ring effect occurs when the first bead of photosphere is uncovered, followed by Baily's Beads and finally the final partial phases of the eclipse.

In the next steps, you will observe the total solar eclipse that occurred on July 11, 1991. This eclipse was notable for its long duration as well as for the fact that the centerline of totality passed over some of the most modern astronomical observatories in the world atop Mauna Kea in Hawaii.

39. Select the view **20N, 155W** under **Go/Observing Projects/Solar Eclipses** and examine the control panel.

The view is centered on the Sun as seen from a location near Hawaii at 7:30:00 A.M. local standard time on July 11, 1991.

40. Manipulate the time flow controls and obtain times for each contact point in the eclipse event.

This sequence, particularly around totality, provides an excellent simulation of a total solar eclipse, including the appearance of bright stars and then fainter stars as the sky darkens, followed by the slow unveiling of the faint corona that can be seen during totality to surround the Sun. While a simulation cannot come close to showing the beauty and majesty of a real eclipse, it nevertheless captures the essence of the event as it unfolds for an observer within the path of totality.

Question 21:   From this location, when do the following events occur?
        a) First contact
        b) Second contact
        c) Third contact
        d) Fourth contact

Question 22:   What is the duration of totality at this location?

The closer a location is to the center line of the eclipse track, the longer the duration of totality.

41. Select the file **19N, 155W** under **Go/Observing Projects/Solar Eclipses**. This position is 1° south of the location in the previous sequence.
42. Manipulate the time controls and obtain the times of each contact of the eclipse event as seen from this location.

Question 23:   What is the duration of totality at this location?

Question 24:   Which location (20N or 19N) is closer to the center line of the shadow track?

The stunning beauty of totality provokes eclipse chasers to try to maximize the time they spend in the shadow of the Moon. Choosing a location as close to the center line of the eclipse track is important, but equally important is to choose the location where the eclipse occurs as close to local midday as possible.

43. Select the view **23N 107W** from **Go/Observing Projects/Solar Eclipses**. Examine the control panel.

The date is July 11, 1991, and this is the same eclipse that you observed from Hawaii in the last two sequences. Note that the local time is near midday at this location in Mexico.

44. Find the times of each contact point of the eclipse as seen from Mexico.

Question 25:   What is the duration of totality at this location?

In the next sequence, you will observe the July 11, 1991, eclipse from a position in space directly over this location in Mexico.

45. Select the view **23N, 107W from Space** under **Go/Observing Projects/Solar Eclipses**.
46. Activate the **Surface Guides** option as described in step 33 if you want to locate the viewing position on the Earth.

In this view, the outline of the Moon's shadow is indicated. Also, a smaller outline within the Moon's shadow represents the umbra of the Moon's shadow.

47. Select a time step of **300×** and activate **Time Forward** or manipulate the time controls to observe the track of the Moon's shadow, particularly that of the umbra, in the July 11, 1991, eclipse.
48. Use **Edit/Undo Time Step** and **Redo Time Step** (or open the view again) to replay the sequence several times.

Question 26: Can you explain the reason for the longer duration of totality when the eclipse occurs near local noon?

F. The Shadow Track

In this section, you will use *Starry Night*™ to examine the last solar eclipse over Europe in the twentieth century, which occurred on August 11, 1999. The eclipse started over the Atlantic Ocean. The shadow then moved quickly over the southern tip of England, over northern France, Germany, Austria, Romania, and out into the Black Sea before covering a strip of Turkey, continuing over Asia, and ending at sunset in the Bay of Bengal on the Indian Ocean. The greatest width of the path of totality was only 112 km, and this occurred over the Black Sea while maximum time of totality was only 2 minutes, 23 seconds, in Romania. First, watch the shadow path of this eclipse as it would be seen from the Moon.

49. Select the view **Aug 11 Eclipse Track** from **Go/Observing Projects/Solar Eclipses**.

Your location is near Bruce Crater on the Moon. The date is August 11, 1999.

50. If you desire, activate the surface guides on the Earth.
51. Start time flow and observe the Moon's shadow move across the Earth.

See step 33 for details on setting surface guides.

Note the small size of the umbral shadow compared to that for the July 11, 1991, eclipse—a fact reflected in the shorter period of totality in the August 11, 1999, eclipse.

You can estimate the speed of the eclipse shadow by timing the simulated eclipse at two different sites, one just south of England, the other in the city of Munich, Germany, some 530 miles or 850 km apart.

52. Select **View from English Channel** under **Go/Observing Projects/Solar Eclipses**.
53. Start the flow of time and record the times of second and third contact.
54. Select **View from English Channel** again under **Go/Observing Projects/Solar Eclipses** and replay the sequence, stopping time flow at second contact.
55. Select **View from Munich** under **Go/Observing Projects/Solar Eclipses**.

56. Keeping in mind the fact that Munich is one time zone East of the English Channel location, adjust the time in the Control Panel so that it is simultaneous with second contact at the English Channel location.

57. Start time flow and determine the times of second and third contact as seen from Munich.

Question 27: What is the time of mid-eclipse (calculated from the measurements of times of 2nd and 3rd contacts) for both sites?

Question 28: Using the two times of central eclipse at these two locations, allowing for the time zone change (subtract 1 hour from your difference), and the distance between the English Channel site and the Munich site of some 530 miles (850 km), what is the speed of the shadow of the Moon as it moves across this area of northern Europe?

G. Conclusions

In this observing project, you have observed solar eclipses from a variety of locations, including above the Earth, sites in Hawaii, the English Channel, Munich, Germany, and the Moon, watching the shadow of the Moon pass over sites while simultaneously watching the eclipse from those sites. You have measured the speed of the shadow of the Moon as it began its passage across Europe and Asia on the last eclipse of the twentieth century over those regions.

PHASES OF VENUS 10

Galileo's discovery of the variable phases of Venus was a turning point in our understanding of the universe. Prior to Galileo's observations, the **geocentric model** (Earth at the center) and the **heliocentric model** (Sun at the center) were simply two competing models. There were no direct observations to allow anyone to differentiate between them. Galileo's observations of Venus provided decisive support for the heliocentric theory.

Most people in Galileo's time and earlier favored the geocentric model, in which the Moon, the Sun, the planets, and all of the stars moved around the Earth. The Moon was the closest of these to the Earth, and the others in order of increasing distance were Mercury, Venus, the Sun, Mars, Jupiter, Saturn, and finally the "sphere of the stars."

The geocentric model of the universe is discussed in Section 4-1 of Freedman and Kaufmann, *Universe*, 7th Ed.

In the simplest geocentric picture, each object moved in a constant direction along a circular path at a constant distance from the Earth. However, this model did not duplicate the observations, particularly the **retrograde motion** of planets in which the normal eastward motion of planets in our sky is interrupted occasionally by westward motion relative to the background stars. To reproduce the observed motions accurately, various complications were added to the geocentric model, most notably by Ptolemy around 140 A.D. The most important of these was that the planet was assumed to move around a smaller circle called the **epicycle**, while the center of the epicycle moved along a larger circle, the **deferent**, around the Earth. This model, when refined, worked very well in predicting planetary motions over short times but became increasingly inaccurate in the centuries following its development.

In the case of Venus, this epicyclic motion brought the planet alternately closer to the Earth and farther away again, but it did not cause Venus to cross the deferent of the Sun. Consequently, in the geocentric theory, even when Venus was farthest from the Earth, it was still closer to us than the Sun. Because Venus never strays far from the Sun in the sky, in this model we should be looking mostly at the dark side of Venus, and we should never see more than a small part of the sunlit side along one edge of the planet. Thus, if the geocentric theory is correct, the apparent size of Venus should change as the planet moves around its epicycle due to its changing distance from the Earth, but we should always see Venus in a crescent phase, or in a "new" phase (entirely dark) when it passes close to the Sun in the sky.

In the heliocentric theory, Venus orbits the Sun. Because its orbit is smaller than that of the Earth, we never see Venus stray far from the Sun in the sky; but it can be anywhere in this orbit as seen from the Earth. If Venus is between the Earth and the Sun, it will show a crescent phase, and at this time it will be closer to us and appear larger. When it is on the opposite side of the Sun from the Earth, we will see it fully sunlit (a full phase), and it will be more distant and therefore appear

Section 4-2 of Freedman and Kaufmann, *Universe*, 7th Ed., discusses the heliocentric model.

Figure 4–14 of Freedman and Kaufmann, *Universe*, 7th Ed., shows how Venus's phases are related to its position in its orbit, as seen from the Earth.

smaller. If the heliocentric theory is correct, we should then see a full range of phases from crescent to full and back again as Venus orbits the Sun, and the phases should be strongly correlated with the apparent size of the planet—largest at crescent phase, and smallest at full phase. It was this observation, along with several other crucial measurements, that convinced Galileo that the planets orbited the Sun.

In this project, you can investigate Venus's motion in the sky and see how this motion correlates with Venus's phases and apparent size.

A. Motion of Venus in Relation to the Sun

We will look toward the Sun from an elevation of 51,025 km above the Earth's North Pole in this simulation. This elevation allows us to watch the behavior of the Sun, stars, and planets over the course of time without either the Moon or the surface of the Earth obstructing our view. We are sufficiently close to the Earth at this altitude that the view of Venus's motion will be equivalent to that seen from the Earth's surface.

1. Launch *Starry Night*™.
2. Select **Venus-A** under **Go/Observing Projects/Phases of Venus**.
3. Open the **View Options** pane, expand the **Solar System** layer and activate the **Labels** option beside **Planets-Moons**.

TIP: If you lose the lock on the Sun, select Edit/Undo Scroll or use Ctrl+Z on the keyboard. Alternatively, select File/Revert and resume time flow.

You should see several planets on the screen. Venus and Mercury are just to the right of the Sun, while Saturn is near the left edge of the screen just below the Pleiades. Jupiter may be just off the left edge of the view and Uranus may be near or just off the right edge.

4. Turn off the **Labels** under **Planets-Moons**, leaving the original labels for the Sun and Venus activated.
5. Run **Time Forward**.

As Venus moves around the Sun, you can see that the Sun also moves relative to the background stars. Because our view is locked on the Sun, the motion of the Sun shows up on the screen as a continuous motion of stars past the Sun. This apparent motion of the stars is an illusion caused by the fact that we are viewing the Sun from a moving Earth. You get a similar illusion if you stand on the edge of a merry-go-round and fix your eyes on an object at the center—it stays fixed in your vision, while the rest of the world seems to spin past it.

Question 1: Approximately how far does Venus move away from the Sun, in degrees, before turning around and coming back toward the Sun? For reference, the total angular width of the screen from one side to the other is 100°.

Question 2: Is the apparent speed of Venus on the screen the same when it is traveling eastward (from right to left) past the Sun as when it is traveling westward (from left to right)? (*N.B.*: Make sure that Time is running forward when you determine the answer to this question)

B. The Ecliptic

You may notice that Venus's motion around the Sun is slightly reminiscent of a figure eight, with a looping path on each side of the Sun. This motion seems to contradict the idea that Venus, as with all planets, orbits the Sun in a single plane.

In fact, the apparent looping motion of Venus is an illusion caused by our present observing position above the Earth's north pole. As the Earth moves along its orbit, our view of the sky moves in response to the tilt of the Earth's rotation axis relative to its orbital plane. From this viewpoint, this causes the plane of the Earth's orbit, the **ecliptic plane,** to appear to sweep through + and $-23^1/_2^\circ$ in the course of a full year. The Sun is always on the ecliptic plane and appears to move along it from our view upon Earth. The edge-on view of this plane, and the apparent path of the Sun across our sky from our view, is known as the **ecliptic.** Because Venus's orbit is almost in the plane of the Earth's orbit, it will also tilt back and forth along with the ecliptic plane.

6. Select **View Options,** expand the **Guides** layer, and activate **The Ecliptic.**

With time progressing forward, you can see that the Sun follows the ecliptic around the sky, and that Venus stays close to the ecliptic (that is, close to the plane of the Earth's orbit) at all times. The small departures of Venus from the ecliptic are caused by the fact that Venus's orbit is inclined by a small angle of about 3° to the Earth's orbit, and therefore to the ecliptic.

7. Activate the **Celestial Grid** in **View Options/Guides.**

The celestial grid (also known as the **equatorial coordinate system**) is a coordinate system on the sky, based upon the Earth's equator and its lines of latitude and longitude. The horizontal lines denote **declination** lines, which are equivalent to lines of latitude on the Earth and are parallel to the equator. We can use these lines as a reference to monitor the change in the tilt of the ecliptic. In this wide-angle view of the sky, the ecliptic appears to be curved. Nevertheless, you can still follow the changing angle of the ecliptic through +/– $23^1/_2^\circ$ with respect to the declination lines as time progresses through the year.

The apparent looping of Venus on each side of the Sun is caused by the change in apparent orientation of the ecliptic (and therefore the apparent orientation of Venus's orbit) relative to this celestial grid in the course of a year.

8. Position the cursor over Venus, open the contextual menu, and activate Venus's orbit.
9. Run **Time Forward** and **Backward** to see Venus following its orbit around the Sun.

C. The Phases of Venus

10. Select **Venus-C** from **Go/Observing Projects/Phases of Venus.**

The view on the screen is similar to that in Part A, except that the view is now locked on to Venus. With Venus locked at the center of the screen, it becomes easy to zoom in to see the phase of Venus, then zoom back out again to see Venus's position and motion relative to the Sun.

11. Run **Time Forward.**

The view on the screen looks different from that in part A because the view is locked on Venus. Consequently, the Sun now appears to orbit around Venus. This is an illusion caused by fixing our gaze on a moving planet. The relative orientation of Venus and the Sun remains the same as in part A, with Venus sometimes to the left and sometimes to the right of the Sun.

12. Select the file **Venus-C** from **Go/Observing Projects/Phases of Venus** to reset the initial view.

In this part of the project, you will observe Venus at various points in its motion relative to the Sun. At each point, you will:
1) Record the date
2) Measure the angular separation between the Sun and Venus
3) Note the distance to Venus from the observing location near the Earth
4) Observe the phase of Venus
5) Determine the angular size of Venus in the sky

Note that, on 4/5/2001, Venus is close to the Sun in the sky.

13. Use the **Hand Tool** to measure the angular separation between the Sun and Venus and note the **Date** and the **Angular Separation** in the Data Table below.
14. Open the **Info** tab for Venus. Under the **Position in Space** layer, find the value given for **Distance from Observer** and record this into the Data Table below, in the **Earth-Venus Distance** column.
15. Zoom in on Venus to a field of view of about 6'.

> **TIP:** If you lose the lock on the Venus, use Edit/Undo Scroll or Ctrl+Z to restore this lock.

You should now see a close-up of Venus, showing its phase. Venus should appear as a very thin crescent.

16. Record the phase of Venus for this date in the Data Table.
17. Under the **Other Data** layer in the Info Pane, find the **Angular Size** of Venus in the sky and record this value into the Data Table.
18. Change the Field of View to 100°.

DATA TABLE

Date	Venus–Sun Separation	Earth–Venus Distance (AU)	Phase of Venus	Angular Size of Venus (arc sec)

In the steps below, you can watch the phase of Venus change as it moves around the Sun.

19. Make sure that the view is still centered (i.e., locked) on Venus and run **Time Forward** to 5/8/2001. Repeat the measurements and observations described in steps 13 to 18 for this date and record your results in the Data Table above.

TIP: An easy way to adjust time one day at a time is to click on the day field of the date in the control panel and use the + and − keys on the keyboard to advance or retreat time.

Question 3: How has the apparent angular separation between Venus and the Sun changed between 4/5/2001 and 5/8/2001?

Question 4: How has the phase of Venus changed between 4/5/2001 and 5/8/2001?

Question 5: Has the apparent size of Venus increased, decreased, or stayed the same from 4/5/2001 to 5/8/2001?

20. Run **Time Forward** to 6/7/2001. Repeat the measurements and observations described in steps 13 to 18 for this date and record your results in the Data Table above.

Venus is west of the Sun (toward the right on the screen) and should now be at about its greatest angular distance from the Sun. This places Venus near **greatest western elongation**.

Question 6: What phase does Venus show when it is near greatest western elongation?

Question 7: Has the apparent size of Venus increased, decreased, or stayed the same from 5/8/2001 to 6/7/2001?

21. Run **Time Forward** to 12/21/2001. Repeat the measurements and observations described in steps 13 to 18 for this date and record your results in the Data Table above.

By 12/21/2001, Venus is close to the Sun again in our sky and moving rapidly toward **conjunction**, the time at which the angular distance between Venus and the Sun is smallest. At this time, Venus is beyond the Sun and at its farthest distance from Earth and hence will appear to be at its smallest at this point.

Question 8: How has the phase of Venus changed between greatest western elongation and conjunction?

Question 9: Has the apparent size of Venus increased, decreased, or stayed the same from greatest western elongation to conjunction?

22. Run **Time Forward** to 8/16/2002. Repeat the measurements and observations described in steps 13 to 18 for this date and record your results in the Data Table above.

By 8/16/2002, Venus is again near its farthest position from the Sun in the sky, this time to the east of the Sun (toward the left on the screen); i.e., Venus is now at **greatest eastern elongation**.

Question 10: How has the phase of Venus changed between conjunction and greatest eastern elongation?

Question 11: Has the apparent size of Venus increased, decreased, or stayed the same from conjunction to greatest eastern elongation?

23. Run **Time Forward** to **10/21/2002**. Repeat the measurements and observations described in steps 13 to 18 for this date and record your results in the Data Table above.

By 10/21/2002, Venus is again approaching conjunction with the Sun, this time from the east.

Question 12: How has the phase of Venus changed between greatest eastern elongation and conjunction?

Question 13: Has the apparent size of Venus increased, decreased, or stayed the same from greatest eastern elongation to conjunction?

The data in the Data Table show you the dependence of angular radius upon the Earth-Venus distance and demonstrates the large change in angular size of Venus during this relative motion of Earth and Venus.

The following steps allow you to see the progression in Venus's phase and apparent size in the sky even more clearly than the steps above.

24. Select **Venus-C** from **Go/Observing Projects/Phases of Venus** to reset the screen to the starting conditions.
25. Set the field of view to 3'.
26. Run **Time Forward,** and allow time to progress until at least 12/20/2002.

Question 14: What relationship is there between the "fullness" of the phase of Venus (e.g., thin crescent, thick crescent, half full, gibbous, full) and the apparent size of Venus?

Question 15: Do your results support the geocentric or the heliocentric theory of the universe? Why?

D. Conclusions

With the observations and measurements in this project, you have demonstrated the large variation in angular radius of Venus that was seen by Galileo almost 400 years ago with his primitive but very effective telescope. It was upon these measurements that he based his conclusion that Copernicus was correct in believing that the planets orbited the Sun rather than the Earth.

PLANETARY MOTION

11

THE RETROGRADE MOTION OF MARS

If we could observe an orbiting planet from a stationary position near the Sun, we would see this planet moving relatively uniformly eastward (i.e., from west to east) against the background sky. There would be small variations in speed caused by the orbit being an ellipse and not a perfect circle, but these variations are relatively small. This eastward motion is called **direct** motion.

In reality, however, we watch a moving planet from a moving Earth, and this leads to a complicated path of the planet across our sky. Consider Mars, for example. Mars is a superior planet because its orbital radius is larger than that of the Earth, and thus it moves more slowly in its orbit than does the Earth. For most of the year we are either on the far side of our orbit from Mars or we are traveling more-or-less toward or away from Mars, and in these cases we see Mars moving "forward" in its orbit; i.e., eastward, or direct motion. To visualize this, imagine yourself in a car heading southward, and ahead of you there is a car on a crossroad traveling eastward. You will see the car traveling from right to left, or eastward, across your line of sight.

However, for part of the year we are on the same side of the Sun as Mars, and overtaking it. Because we are pulling ahead, Mars appears to drop back in a westward direction as seen from the Earth, just as the other car would appear to drop back relative to us if we turned onto the crossroad and started to pass it. This temporary westward motion is called **retrograde** motion. For the same reason, we see retrograde motion for the other superior planets.

> Opposition, conjunction and other planetary configurations, retrograde motion, and sidereal and synodic periods are discussed in Section 4–2 of Freedman and Kaufmann, *Universe*, 7th Ed.

> *The Synodic Period of Mars* project explores this difference between the sidereal and synodic periods of Mars.

You can demonstrate this motion easily with *Starry Night*™, by watching the relative motion of Mars night by night against the background stars. You will find that the reversal of apparent motion takes place when the planet is at **opposition,** on the opposite side of the Earth from the Sun. This is expected from the argument above, because it is at opposition that the Earth is overtaking the more distant and hence slower planet.

The time taken for the Earth-Mars system to move from one opposition to the next is known as the **synodic period** of Mars. This differs from the more fundamental revolutionary period of Mars with respect to the stars, its **sidereal period**. The latter cannot be measured directly from Earth but its value can be determined from a simple calculation after measurement of the synodic period, as is shown later in this project.

A. Mars Motion in the Sky

The particular period chosen for this simulation includes one of the closest approaches of Mars to Earth in the past decade as these planets follow their respective elliptical orbits. (In fact, in this decade, only at the 2003 opposition, one synodic period before the one observed in this simulation, would Mars be closer to the Earth. As an extension after exploring this project, you might like to demonstrate that this is in fact the case, particularly with the view from space of the relative orbits of Earth and Mars in Section C.)

We will begin by observing the motion of Mars as seen from the Earth and explore the change in angular size of Mars as it moves through its close approach.

   1. Launch *Starry Night*™.
   2. Select **File/Preferences/Cursor Tracking (HUD)** and choose only **Name** and **Object type** from the **Show** list.
   3. Select the view **Mars-A** from **Go/Observing Projects/Planetary Motion**.

In the early morning hours of August 19, 2005, in Montreal, Canada, it is already twilight and the sky appears dark blue. Mars is faintly visible near the center of the screen. Relative to the horizon, it is almost due south. The following steps allow you to verify this.

   4. Identify Mars with the **Hand Tool**.
   5. Drag the screen up to show the compass points along the horizon, and check the position of Mars relative to the south point.
   6. Select **Mars-A** from **Go/Observing Projects/Planetary Motion** to reset the view.

To demonstrate the complex motions of Mars in our sky, we can run time forward in sidereal-day steps to maintain the same background of stars against which Mars's progress can be followed. For this initial investigation, we will leave daylight on to show the real appearance of the sky through the next few months, until Mars moves into evening twilight. Later, we will turn off the daylight to follow Mars's progression over a longer period.

   7. Run **Time Forward** to 2/05/2006 A.D.

If you watch the time display, you will notice that the time of night becomes earlier and earlier as the days pass. This is because 1 sidereal day is about 4 minutes short of 1 solar day. Thus, as the days pass, you see morning twilight deepen into night and then eventually brighten into evening twilight. By February 5, 2006, the sidereal day steps have moved the observing time to 5:32:39 P.M. and Mars is beginning to fade into evening twilight. Mars can be seen just under the quarter moon on this evening.

Mars's motion is initially direct motion to the east, but it soon reaches a **stationary point** where it stops and reverses its direction. It then moves westward for a period of retrograde motion. Eventually, as evening twilight begins to brighten the sky, Mars reaches a second stationary point where its retrograde motion ceases and direct or easterly motion resumes.

   8. Select the view **Mars-A** from **Go/Observing Projects/Planetary Motion** to reset the view.
   9. Run the time sequence again, stopping **Time** at each stationary point to note the **date** and answer the following questions.

   Question 1:   On what dates is Mars at its stationary points during this retrograde period?

10. In order to answer Question 2, open the **Find** pane and expand the Moon layer to check that the option **Enlarge Moon size** is activated. Click **OK**.
11. Use the time controls to move the date in sidereal day steps to **November 14, 2005,** near the midpoint between the two stationary points, on a day when the Moon is near Mars in the sky. Note particularly the phase of the Moon. What does this tell you about the position of the Sun relative to Mars at this time and date?

Question 2:   Where would the Sun be at this time, relative to the Earth and Mars?

Question 3:   What would be the configuration of Mars at this time (e.g., conjunction, maximum elongation, opposition, etc.)?

In this simulation, you advanced time in sidereal-day steps to maintain the constant background of stars. This led to a uniform shift in the time of observation every night, such that twilight and eventually daylight rendered Mars invisible at either end of the observing period. This limited your view of the motion of Mars to a short period around its retrograde phase. In *Starry Night*™, you can artificially turn off the daylight and extend the run time to observe the apparent motion of Mars over a longer period.

12. Select the view **Mars-A2** under **Go/Observing Projects/Planetary Motion.**

In this view, in June 2005, daylight has been turned off. Mars and the Sun are labeled. Notice that the Sun is to the east of Mars in the sky. Knowing that the default screen width is about 100°, you can see that the Sun and Mars are about 90° apart in the sky.

13. Run **Time Forward** in sidereal day steps and stop **Time** when the Sun appears in the western sky, on or about March 24, 2006 (3/24/2006 A.D.).

You can see from this long run, uninterrupted by daylight, that a long period of direct motion gives way to retrograde motion for a relatively short period between May and July.

B. Phase and Angular Size of Mars

The relationship between the synodic and sidereal periods of a planet is discussed in Box 4–1 of Freedman and Kaufmann, *Universe*, 7th Ed.

In the following simulation, we will view Mars from an elevation of 12,756 km above the Earth's North Pole. This elevation is high enough to allow us to watch Mars all day without the surface of the Earth obstructing our view but we are still sufficiently close to the Earth that our view is equivalent to that seen from the Earth's surface. We are also outside the Earth's atmosphere, so that our view is not blocked by the blue scattered sunlight of daytime. In the absence of daylight, we can examine the appearance and angular size of Mars as it moves through one synodic period, starting from conjunction with the Sun and progressing through opposition to the next conjunction.

Mars is a superior planet with an orbit larger than that of Earth. Thus, at conjunction this planet is beyond the Sun, whereas at opposition it is much closer to us. It therefore shows a varying size and brightness over its synodic period. Also, because it never passes between the Earth and the Sun, it will not display the full range of phases seen in Mercury, Venus, or the Moon. In fact, as we shall see, its appearance changes only slightly, from a full phase to a gibbous phase with only a small region of shadow, in the course of its orbital motion.

14. Select the view **Mars-B** from **Go/Observing Projects/ Planetary Motion.**

The view is centered on Mars, which appears just above the Sun on September 15, 2004, very close to conjunction. By starting at this position, you can determine Mars's synodic period by measuring the time needed to return to this configuration again.

TIP: If you lose the lock on Mars, use **Edit/Undo Scroll** or use the keyboard shortcut **Ctrl +Z.**

This view is centered upon Mars. Thus, when you advance time, the background stars, the planets and the Sun move in various ways while Mars remains stationary. This mode should be maintained throughout this sequence.

In the following part of the project, you will observe Mars at various points in its motion relative to the Sun. At each point, you will:

1) Record the date
2) Note the distance of Mars from the Earth
3) Observe the phase of Mars
4) Determine the angular size of Mars in the sky

15. Record the **Date** in the Data Table below.
16. Open the **Info** tab for Mars. Under the **Position in Space** layer, find the value given for **Distance from Observer** and record this into the Data Table below, in the **Distance to Mars** column.
17. Zoom in on Mars to a field of view of about 1 arc minute using the slide bar and the zoom button. It is useful to zoom in to about the same field of view on each successive observation in order to watch the change in apparent size of Mars during the following sequence.

If the top layer of the Info pane does not say "Mars Info," then right-click on the planet Mars in the main view and click Select **Mars** in the contextual menu.

In this expanded view, you will see Mars and its two moons, Phobos and Deimos. Note the detail on Mars's surface. These features will become more distinct as Mars comes closer to Earth.

18. Under the **Other Data** layer in the Info Pane, find the **Angular Size** of Mars in the sky and record this value into the Data Table.
19. Record the phase of Mars for this date in the Data Table.
20. Slide the Field of View back to 100°.
21. Make sure that the view is still centered (i.e., locked) on Mars.

Configuration	Date	Distance to Mars	Mars's Angular Size	Phase
Conjunction	9/15/2004			
Stationary point				
Opposition	11/05/2005			
Stationary point				
Conjunction				

You can repeat the above measurement sequence twice before Mars reaches its first stationary point, in order to obtain representative values of distance and angular radius of Mars as seen from Earth.

22. Run **Time Forward** for four months to **1/15/2005** A.D. Repeat the measurements and observations described in steps 15 to 21 for this date and record your results in the Data Table above.
23. Run **Time Forward** another 4 months to **5/15/2005** A.D. Repeat the measurements and observations described in steps 15 to 21 for this date and record your results in the Data Table above.

The next three measurements should be made at stationary points and at opposition. You need to run time forward until the background appears to stop moving behind Mars, indicating that Mars has reached its first stationary point. You should use single time steps to adjust Mars to this position.

24. Run **Time Forward** until Mars stops moving eastward against the background stars at the **first stationary point**. Repeat the measurements and observations described in steps 15 to 21 for this date and record your results in the Data Table above.
25. Run **Time Forward** to **November 6, 2005**, when Mars is at **opposition**. Repeat the measurements and observations described in steps 15 to 21 for this date and record your results in the Data Table above.
26. Run **Time Forward** until Mars reaches the **second stationary point**. Repeat the measurements and observations described in steps 15 to 21 for this date and record your results in the Data Table above.
27. Take two further measurements at about six-month intervals.
28. Finally, run **Time Forward** until Mars reaches **conjunction** again and is near to the Sun in the sky. Repeat the measurements and observations described in steps 15 to 21 for this date and record your results in the Data Table above.

Question 4:   How long does Mars's retrograde period last?

Question 5:   What is the synodic period of Mars?

Question 6:   How much bigger does Mars appear to us at opposition compared to its appearance at the previous conjunction?

It is instructive to rerun this time sequence at high magnification without stopping, and observe how the phase of Mars changes from full to gibbous and back again to full while its angular size also changes. The moons of Mars will be moving rapidly around the planet as this simulation proceeds. Watch particularly how the spin axis of Mars tilts to reveal more or less of the polar ice caps of this planet. Note also how the dark shadow edge, which makes Mars appear gibbous, moves from one side of Mars to the other as this motion proceeds. It is instructive to run the sequence several times to watch this impressive simulation of the varying appearance of Mars as viewed from a moving Earth.

29. Select the view **Mars-B** from **Go/Observing Projects/ Planetary Motion.**
30. Set the **Field of View** to about 1'.
31. Run **Time Forward** to watch the changes in the appearance of Mars as the planet proceeds through one synodic period from conjunction to conjunction.

In this simulation, Mars appears to rotate in the "wrong" direction, counter to its orbital direction. This is caused by the fact that we are advancing time in steps of 24-hour days, whereas Mars' true rotation period is about $24\frac{1}{2}$ hours. Thus, Mars will appear to have rotated by a little less than one rotation in every Time Step, and this stroboscopic effect produces an apparent "retrograde" rotation of Mars.

C. Orbital Motion of Mars from Space
Starry Night™ can display the planets from a position out in space and show their orbital motions. This will demonstrate to you the actual positions of Mars and the Earth during the previous sequence.

32. Select **Mars-C** under **Go/Observing Projects/ Planetary Motion.**

This view shows the inner solar system from above the north pole of the Sun. On this date, September 15, 2004, Mars is on the opposite side of the Sun from the Earth, and therefore in conjunction with the Sun as seen from Earth.

33. Run **Time Forward** and watch Mars move with respect to the Earth, noting particularly the dates of opposition and of the next conjunction.

From this unique vantage point at the "ecliptic pole," the view is directly down upon the planets moving around in their respective orbits. You can see how the elliptical orbits of Earth and Mars can bring Mars closer to Earth at some times and not at others, depending upon their positions in these orbits. You can experiment with the settings to demonstrate this fact more clearly, for example in the view Mars-C2 under Go/Observing Projects/Planetary Motion. You might also select a preferred time for observing Mars in history or in the future.

D. Conclusions
You have followed Mars as it orbits the Sun, watching it from a moving platform, the Earth, and have seen it move from direct to retrograde motion as Earth catches up the slower-moving Mars in its larger orbit. You have measured the change in angular size of Mars that results from this relative motion of Earth and the planet. Finally, you have observed the "real" motion of these planets from above the solar system. This relatively uniform motion around the Sun cannot be seen from the moving Earth.

MARS AND ITS MOONS

12

The planet Mars has fascinated people throughout history. Its reddish color and regular brightenings over about a 2-year cycle were considered to be significant and auspicious, and this "wandering star" was regarded as a god of war by many civilizations. Later, misinterpreted descriptions of features on its surface led to the belief that Martian inhabitants had constructed canals across its surface. An extensive science fiction literature developed on these themes, and only now, after several sophisticated spacecraft have orbited and landed upon Mars's surface, has this flurry of speculation given way to a more sober examination of the possibility of elementary life on this neighboring planet.

We will use *Starry Night*™ to look briefly at the surface of Mars, and then take a closer look at the two tiny moons, Phobos and Deimos, discovered in 1877 by the American astronomer Asaph Hall. They were named after the mythical horses that pulled the chariot of Mars, the god of war. They are irregular in shape, have average diameters comparable to that of a small city, and move in almost circular orbits at relatively low altitudes above the planet's equator.

Phobos, its name meaning fear, is about twice the size of Deimos. Its period is much shorter than the rotational period of Mars, and as a consequence, as seen from the surface of Mars, Phobos rises in the west and moves rapidly across the sky to set in the east several hours later.

Deimos, its name meaning panic, orbits at a much larger distance from the planet than Phobos and hence its orbital period is longer. In fact, Deimos's orbital motion is almost synchronous with the rotation of Mars itself and so Deimos creeps slowly across the planet's sky, taking about 3 Martian days to go from one horizon to the other.

Historically, it is interesting to note that in 1726, 150 years before the discovery of these satellites of Mars, Jonathan Swift, in his satirical novel *Gulliver's Travels*, wrote of astronomers in the land of Laputa who had discovered two satellites that revolved around Mars! Their orbital properties were described as follows:

> *"They have likewise discovered two lesser stars, or satellites, which revolve about Mars, whereof the innermost is distant from the center of the primary planet exactly three of its diameters, and the outermost five; the former revolves in the space of ten hours, and the latter in twenty-one and a half; so that the squares of their periodical times are very near the same proportion with the cubes of their distance from the center of Mars, which evidently shows them to be governed by the same law of gravitation that influences the other heavenly bodies."*

The references to Kepler's third law of planetary motion and Newton's law of gravitation indicate that Swift, who wrote the book in the latter years of Newton's life but was not a scientist, nevertheless understood this "latest scientific theory" of gravitation. It is difficult to know what to make of his "prediction" of the existence of the moons of Mars, 150 years before their discovery! You will be able to use *Starry Night*™ to compare the actual parameters of the moons' orbits with these predictions.

A. Mars: Surface Features and Rotation

The geometry of Mars's orbit and its apparent motion in our sky, including stationary points, is explored in the Planetary Motion project and discussed in Section 4–1 and shown in Figure 4–2 in Freedman and Kaufmann, *Universe*, 7th Ed.

In order to observe the planet for a reasonable period without interference from the horizon or daylight, the viewing location will be at the Earth's North Pole in winter during a period when Mars is relatively close to Earth and at a "stationary point." At a stationary point, Mars appears to stop and reverse its apparent eastward movement in our sky for a specific period.

1. Launch *Starry Night*™.
2. Select **Mars From North Pole** under **Go/Observing Projects/Mars and its Moons.**

Hint: Open the **Find** pane, expand the Moon layer, and select the option **Enlarge Moon size.**

You might take a moment at this stage to orient yourself to the location and to look around the sky. At this time of perpetual winter at the North Pole, Mars appears about 16° above the southern horizon, just to the east of a gibbous Moon. Also visible in the southeast is the open cluster of stars, the Pleiades, while the spiral galaxy in Andromeda can be seen high in the sky, just to the west of south. You can use the Hand Tool to identify these features of the night sky, move them to center-screen and zoom in on them to see their true beauty. If you scroll the view to break the lock on Mars and run time forward, you will see the stars, the Moon, and Mars moving parallel to the horizon from this northernmost viewpoint as the Earth rotates around the North Celestial Pole. (The cardinal points of the compass are displayed around the horizon, but they mean little at this location because all directions from here are south!)

3. After exploring this sky, select **Mars From North Pole** again under **Go/Observing Projects/Mars and its Moons.**

We can now examine the surface of Mars briefly and measure its rotation period.

4. Change the **Field of View** to about **3 arc minutes.**

You can compare the overall animated view presented with *Starry Night*™ with the more detailed images of Mars shown in Chapter 13 of Freedman and Kaufmann, *Universe*, 7th Ed., and with the videos presented on the compact disks accompanying the book.

This view shows Mars with its two tiny moons, and allows you to judge the scale of the moons' orbits compared to the planet's radius. You will measure these parameters later in this project. You can now zoom in to fill the screen with the planet, to examine some of its features.

5. Zoom in to a **Field of View** of about **33 arc seconds.**
6. Select **File/Preferences...** from the menu. In the **Cursor Tracking (HUD)** page, choose only **Name, Object type,** and **Surface feature** from the **Show** list. This latter feature will allow you to identify features on Mars's surface with the **Hand Tool.**

The rotation axis of Mars at the start of this run is inclined at about 45° to the vertical on the screen. This initial view shows many dark features in the lower half of the planet below a brighter, less featured region.

7. Select **3000x** in the time step control and watch Mars rotate slowly around its axis, stopping time flow occasionally to use the **Hand Tool** to identify some of the features in the following descriptions.

As rotation proceeds, you will discover that Mars has two distinct hemispheres. (The moons will flash by in their orbits in this expanded view. We will examine these motions later.) Around 02:00:00 A.M., a much darker region with considerable structure appears and leads to another very dark area extending well into the northern hemisphere, an area known as Syrtis Major. These dark regions are well seen by 04:00:00 A.M.

Around 11:30:00 A.M., just as Syrtis Major is disappearing, a set of extensive canyons begins to appear. By 4:00:00 P.M., these canyons can be seen to begin near a line of three large volcanoes, and extend from there a significant way around the planet.

Question 1: What are the names of these three volcanoes?

Near 5:30:00 P.M., a fourth massive volcano, Olympus Mons, has appeared. This volcano has a height about three times that of Mt. Everest and a base that extends over a distance of 600 km.

The tilt of the axis of Mars is such that only a very small polar cap can be seen in the southern hemisphere at this time. You will use a different viewpoint to look at these regions later in the project.

In the next sequence, you will measure the rotation rate of Mars.

8. Select **Close-up of Mars** under **Go/Observing Projects/Mars and its Moons.**

The view shows a close-up of Mars from the North Pole of Earth. The red line across the face of Mars shows the planet's meridian. The time has been adjusted so that this line is straight.

9. Note the date and time of this observation in Data Table 1 below in the row for Event #1.
10. With the **time step** at one minute, start **Time Flow Forward** until the meridian line returns to the same position and appears straight across the face of the planet. Adjust the time with the single step controls to make the meridian line precisely straight. Note the date and time of this observation in the row for Event #2 in Data Table 1 below.
11. Obtain the rotation period of Mars in hours and minutes by finding the time difference between Events #1 and #2.

TIP: To adjust the time so that the meridian line is precisely straight, hold a piece of paper lightly against the screen along the meridian line.

DATA TABLE 1

	Mars Rotational Period	
	Date	Time
Event #1		
Event #2		
Difference		

The difference obtained in the Data Table above actually represents the synodic rotation period of Mars as measured from a moving platform, the Earth. The required correction to allow for the movement of Earth relative to Mars is small in this case and your value can be compared to the quoted (sidereal) rotation period of this planet.

Question 2:   What is the rotation period of Mars, in hours and minutes?

B. Mars's Moons from Earth

We can observe the moons of Mars from Earth and measure their orbital parameters from a view equivalent to that of an Earth-bound telescope. From these observations, we can check the "predictions" made by Swift's astronomers in *Gulliver's Travels.*

12. Select the file named **Mars and Orbits of Moons** under **Go/Observing Projects/Mars and its Moons.**
13. Start **Time Flow Forward** and observe the motions of the moons as they orbit Mars.

The moons move around the planet in the equatorial plane of Mars. The fact that the Moons move in ellipses on the screen, rather than following a straight line back and forth across Mars, shows that at this time their orbital planes are inclined away from the Earth.

As time advances, you will see that Phobos appears to pass behind Mars. This type of event, where an object is obscured by another object, is known as an **occultation.** Disappearance into an occultation is known as **ingress,** while reappearance is known as **egress.** In order to answer the following question, you should watch very carefully the reappearances of Phobos from behind the planet and the appearance of Deimos as it moves beyond the planet on every orbit.

Question 3:   Why does Phobos appear very faint or invisible when it emerges beyond the planet's limb, and what causes Deimos to appear to wink out at a certain position just beyond the planet, on every one of their orbits?

Phobos will also appear to pass in front of the planet in an event known as a **transit.** During a transit, you will note that Phobos overtakes features on the surface, indicating that, as viewed from the surface, this moon would appear to move in the same direction as the rotation of the planet; i.e., counter to the apparent sky rotation. Single 1-minute time steps and a larger magnification at the appropriate time will show this relative motion more clearly. From this view, you can see how close Phobos is to the planet's surface in its orbit. In contrast, Deimos moves much more slowly in its larger orbit around the planet.

We can now determine the orbital parameters of radius and period for each moon around Mars, assuming that the orbits are closely circular, which is in fact the case.

14. Stop **Time Flow.**
15. Use the **Hand Tool** to measure the angular radius of Mars, in seconds of arc, and enter the value in Data Table 2 below. (If in this or the following two steps the angle is more than 1 minute of arc, multiply the arc minutes by 60 and add this to the number of arc seconds to find the angle in seconds of arc.)

16. Measure the orbital distance of Phobos (the angular distance from the center of Mars to either "end" of Phobos' orbit, as seen on the screen) in seconds of arc, and enter the value in Data Table 2.
17. In similar fashion measure the orbital distance of Deimos in seconds of arc, and enter the value in Data Table 2.

DATA TABLE 2

Measurement	Angle (")	Ratio of Orbit to Planet Radius
Mars radius		1
Phobos orbital radius		
Deimos orbital radius		

Question 4:   By what factor is Phobos's orbit bigger than Mars's radius?

Question 5:   By what factor is Deimos's orbit bigger than Mars's orbit?

You can now test the "predictions" of Swift's astronomers of Laputa.

18. Calculate the orbital distances of Phobos and Deimos in kilometers, using the ratios from Data Table 2 above and the radius of Mars (3397 km), and enter the values in Data Table 3 below.
19. Calculate the ratios of these orbital distances to the diameter of Mars (6794 km), and enter the values in Data Table 3.

DATA TABLE 3

Swift's *Gulliver's Travels* Prediction Orbital radius as a ratio of Mars's diameter (6794 km)			
Moon	Orbital Distance in km	Ratio of Orbital Radius to Mars's Diameter	Swift's Ratios
Phobos			3.0
Deimos			5.0

To measure the orbital period of a moon, we need to measure the time taken for the moon to pass completely around the planet, using a reference point for timing this motion. The best reference point for Phobos is the passage of this moon across the edge of Mars's image at the beginning of a transit. (Again, this measurement will be in error because it is carried out from a moving Earth. In practice, the correction for this motion is small and can be ignored.)

20. Change the **time step** to **1 minute**.
21. Run **Time Forward** until Phobos is crossing a limb of Mars, adjusting the time in single-step mode to place it as close as possible to this position. Note the date and time in Data Table 4 below, in the row for Time No. 1.
22. Run **Time Forward** until Phobos returns to the same position one orbit later and note the date and time in Data Table 4, in the row for Time No. 2.

In the case of Deimos, this moon does not cross the planet's limb on this date as seen from the Earth. A convenient reference point is the time when it is passing over the spin axis of the planet.

23. Open the **View Options** pane, expand the **Solar System** layer, and choose **Planets-Moons Options...** by positioning the cursor over **Planets-Moons**.
24. In the Planet Options dialog, activate **Surface guides** and choose only **Pole sticks** from the sub-options. Click **OK**.
25. Use the time controls to position Deimos over the pole stick at Mars's north pole. Note the date and time in Data Table 4 below, in the row for Time No. 1.
26. Run **Time Forward** until Deimos returns to this position over the north pole stick. Use single steps in time to adjust Deimos's position as precisely as possible and note the date and time in Data Table 4, in the row for Time No. 2. *(Hint:* It might be easier to step time in hours, minutes, and seconds while counting these steps to determine this time, thereby avoiding the somewhat awkward subtraction of two times.)
27. The differences in these times will be the orbital periods of the moons, for comparison with Swift's predictions.

DATA TABLE 4

	Phobos	**Deimos**
Time No. 1		
Time No. 2		
Difference in Times (Orbital Periods)		
Swift's Predicted Periods	10 hours	21.5 hours

Question 6: What is the orbital period of Phobos around Mars?

Question 7: How does this period compare with the rotation period of Mars?

Question 8: What is the orbital period of Deimos?

Question 9: How close were the predictions of Mars's orbital distances and periods by Swift's astronomers?

C. Mars's Moons from the Martian Surface

Because Phobos and Deimos have much shorter orbital periods than our own Moon, it is fascinating to observe their motions as seen from the surface of Mars. One day, space explorers might experience these effects in real life. Until then, we can use *Starry Night*™ to simulate the motions of these moons!

The brightnesses of Phobos and Deimos are very low compared to our Moon because of their small sizes. In the Martian sky, Phobos is several times brighter than is Venus in our sky while Deimos is about as bright as Venus. Both moons move around Mars in the direction of the planet's rotation, but the orbital period of the inner moon, Phobos, is so short that it moves faster than the Martian surface. Thus, as seen by someone on the surface of Mars, it will appear to move rapidly across the Martian sky from west to east and in fact will cross this sky twice per Martian day. The orbital distance of the outer moon, Deimos, is such that it almost keeps pace with the Martian surface and is seen to move very slowly from east to west across the sky, taking more than 5 Martian days to complete a full orbit with respect to the planet's surface. In this respect, therefore, it almost mimics the synchronous communication satellites in our sky, whose orbital periods accurately match the rotation period of the Earth.

You will observe these effects in the following sequence.

28. Select **View from Mars Surface** under **Go/Observing Projects/Mars and its Moons.**

In fact, this is not the location of the landing site of this spacecraft. This scene appears in the northwest in all surface simulations of Mars. The location of the actual landing site, at latitude 15°N and longitude 330°E on Mars, is shown on the map in Figure 13–5 of Freedman and Kaufmann, *Universe*, 7th Ed.

Note that Time is specified in Universal Time. At this time and location, it is dark on Mars. If you scroll the view around the horizon, you can see the Mars Pathfinder rover "Sojourner" among several rocks that appear toward the NW in this simulation. If you lower your view direction, you will see the spacecraft from which this rover emerged after landing on Mars in July 1997.

You may recognize several of the constellations in this sky. You can toggle on the constellation patterns using the K key to identify Leo the Lion just above the horizon in the southeast, Ursa Major high in the eastern sky, and Gemini, the Twins, high in the southern sky.

29. Toggle on the display of constellations using the K key. Press the K key once more to turn the constellation display off again.
30. Select **View from Mars Surface** again under **Go/Observing Projects/Mars and Its Moons** once again to return to the initial configuration. Select a time step of 300× the rate of real time and watch the sky from Mars, paying particular attention to the motion of the moons Phobos and Deimos.

The Moon's motion in our own sky as seen from the surface of the Earth is demonstrated in the Moon's Motion and Phases project.

As on Earth, the stars rise in the eastern (or here the southeastern) sky and move toward the west because of the rotation of Mars. Meanwhile, Phobos moves quickly in the opposite direction, from west to east, and soon sets below the same horizon from which the stars are rising. You might notice that the motion of Phobos relative to the background stars is in the same direction (west to east) as that of our own Moon as seen from the surface of the Earth. However, Phobos's orbital motion is so fast that it actually moves from east to west relative to the horizon, whereas our Moon's orbital motion is so slow that the Earth's rotation causes it to drift westward along with the stars, relative to the horizon.

Deimos appears almost to hang in one spot while the stars move past it. This is an illusion created by the rotation of Mars. In fact, it is really Deimos that is moving toward the east past the stars at a rate that almost exactly keeps pace with the rotational motion of Mars. However, its orbit is not quite synchronous with Mars's rotation: if you watch closely, you may see that Deimos moves very slowly toward the west relative to the horizon. It takes just under 3 days to cross the Martian sky from horizon to horizon, or about 5.5 days to return to the same point in the sky again, relative to the horizon.

At about 18:45:00 UT, just as the landscape begins to brighten before sunrise, Jupiter and the Earth rise almost simultaneously above the horizon. A little over an hour later the Sun rises into a typical Martian sky. The light reddish-brown color is caused by scattering of sunlight from the very fine dust blown into the atmosphere by frequent windstorms. It is well worth re-running this spectacular simulation of a Martian sunrise!

It is also interesting to look at Earth from Mars as the Sun rises, with the field of view of a telescope, much as future explorers might watch the home planet from their landing site on Mars.

31. Stop time flow and adjust the Time to **19:00:00 UT** on **12/31/2005**.
32. Open the **Find** pane and click on **Earth** to **Center** the view upon the Earth.
33. Select **View Options/Solar System** and position the cursor over **Planets-Moons** to open the **Planets-Moons Options**. Activate the **Show atmosphere** option and click **OK**.
34. Change the field of view to **1 arc minute**.
35. Select a time step of **300×** and watch the blue, white, and green Earth slowly fade as the sky brightens with sunrise on Mars.

Question 10: What is the phase of the Earth at this time?

D. Mars and its Moons from Above

We can view the orbits of Mars's moons from above the plane of the ecliptic and watch the moons from a fixed reference point so that they are seen to move around the planet with their true sidereal periods.

36. Select **Mars and Moons from Space** under **Go/Observing Projects/Mars and its Moons**.
37. Select a time step of **3000×** the rate of real time and watch the moons move around the planet.

If you look closely, you can see that Mars is also rotating in the same direction as the motion of the two moons. It is interesting to pick a Martian surface feature that is on the same side of Mars as Deimos, and watch this feature and Deimos as time passes. You might see that the relative orientation of the feature and Deimos remains almost constant, showing that Deimos's orbit is almost synchronous with Mars's rotation.

From this unique view, you can determine the sidereal periods directly by timing the passage of these moons past some reference star.

38. For each of Mars's moons, choose a reference star very near its orbital path.
39. Change the **time step** to **1 minute** and adjust the time with the time controls so that the moon is on the line joining the star to the center of Mars.
40. Record the date and time of Passage 1 for the moon in Data Table 5 below.
41. Run **Time Forward** and, when the selected moon approaches the star again, use the single step controls to position the moon in the same direction as the star from the center of Mars.
42. Record the date and time of Passage 2 in Data Table 5 below.
43. Calculate the differences in the times of these events to determine the sidereal period of each moon.

DATA TABLE 5

	Phobos		Deimos	
Event	Date	Time	Date	Time
Passage 2				
Passage 1				
Difference				

Question 11: What is the sidereal period of Phobos, expressed in days?

Question 12: What is the sidereal period of Deimos, expressed in days?

From this viewpoint, you can take a closer look at the polar cap of Mars by zooming in until the Field of View is about 5°. Because this northern polar cap is more easily visible when it is fully sunlit, you should change the date to October 1, 2006 (10/01/2006 A.D.). This image is derived from a real Mars photograph, but the radial structure at the extreme pole is caused by the extreme angle from which it was taken by spacecraft that orbit near to the planet's equator. Nevertheless, this simulation provides a magnificent view of this remote planet and its icy polar cap.

E. Eclipse of Deimos by Phobos

Unlike the Earth, Mars has two moons, so it is possible for an observer on Mars to see the inner moon, Phobos, pass in front of, or eclipse, the outer moon, Deimos. In this section you will be able to watch a simulation of such an eclipse.

The orbits of Phobos and Deimos are inclined relative to Mars's equator by 1.08° and 1.79°, respectively. An observer close to Mars's equator would then see Phobos's and Deimos's orbits inclined relative to each other by about 0.7°, and would therefore see the orbits intersect at two points in the sky, much as we on Earth see the ecliptic and the Moon's orbit intersect at two points. If Deimos is near one of these intersection points when Phobos passes by, the observer would see an eclipse of Deimos by Phobos.

44. Select **Eclipse-Wide** under **Go/Observing Projects/Mars and its Moons.**

The view is toward the east from the surface of Mars on December 31, 2005, at a location 2° south of the Martian equator. Phobos and Deimos are visible near the center of the screen, flanked by the Big Dipper on the left and Leo the Lion on the right.

45. Select a time step of 300× and watch how the moons move.

As we saw earlier, Phobos moves quickly toward the east (down, in this view), opposite to the apparent motion of the stars while, over the short time represented on the screen, Deimos remains almost stationary above the horizon. At about 15:30 UT, Phobos eclipses Deimos.

46. Select the file **Eclipse-Telescopic** under **Go/Observing Projects/Mars and its Moons.**

The screen now shows a telescopic, or magnified, view of Deimos at a time close to the eclipse you watched in the previous sequence.

47. Select a time step of 30×.

After a few seconds, Phobos passes rapidly through the area of sky visible on the screen, eclipsing Deimos as it does so.

48. Select **Eclipse-Telescopic** again under **Go/Observing Projects/Mars and its Moons.**
49. Run **Time Forward,** and then stop Time when Phobos is near the upper edge of the screen.
50. Display the orbits of both Phobos and Deimos.
51. Change the Time Step to 30× and watch the eclipse again.

You can see that Deimos is indeed near an intersection point of the two orbits. You can also see that, because Phobos appears much larger than Deimos in the sky, and because the orbits are inclined at such a small angle relative to each other, Deimos can actually be some distance away from the intersection point and still be eclipsed by Phobos.

F. Conclusions

The appearance and rotation of Mars and the revolution of its two small moons have been examined from various novel viewpoints, including the surface of the planet itself. These observations were used to measure the radii and periods of the orbits of the moons. These values were then compared to the fictitious "predictions," made by Jonathan Swift in his novel *Gulliver's Travels* 150 years before the moons were actually discovered! You have also viewed the eclipse of one moon by the other, a phenomenon impossible on Earth because our planet has only one Moon.

SYNODIC PERIOD OF MARS 13

The **synodic period** of a planet is the time taken for the planet to travel from any particular planetary configuration as seen from the Earth to the next occurrence of the same configuration; e.g., for a superior planet it could be the time from conjunction to the next conjunction.

The **sidereal period** of a planet is the time taken for the planet to orbit the Sun once, relative to the distant stars.

Synodic and sidereal periods differ in viewpoint. For sidereal period we can think of a hypothetical observer at rest on the Sun. If this observer sees the planet line up with a par-

| Section 4–2 and Box 4–1, Tools of the Astronomer's Trade in Freedman and Kaufmann, *Universe*, 7th Ed., discusses planetary configurations and synodic and sidereal periods. |

ticular star today, then one sidereal period later the planet will line up again with the same star. Alternatively, we can think of a hypothetical observer at rest above the solar system, looking down on the Sun and the planets. If this observer imagines a line drawn from the Sun through the planet to a particular star, then this planet will move through exactly 360° and return to this line again in one sidereal period. For the synodic period we are the observers, watching the solar system from a moving Earth and thus the period we measure for a planet relative to the Sun depends on the motion of the planet and the motion of the Earth.

In this project you can use *Starry Night*™ to see how synodic periods arise, and how and why they differ from sidereal periods.

A. The Synodic Period of Mars

> 1. Launch *Starry Night*™.
> 2. Select **Synodic-A** under **Go/Observing Projects/Synodic Period of Mars**.

The view shows the sky from London, England, at 10 P.M. on August 10, 2002. Mars is in conjunction with the Sun, meaning that Mars is beyond the Sun as seen from the Earth. Mars's orbit is inclined slightly compared to the Earth's orbit so we see Mars just above the Sun in the sky, rather than directly behind it.

The green line represents the ecliptic, or the plane of the Earth's orbit. The ground and horizon have been made invisible as if we were observing the sky from a completely transparent Earth so that, when you run time forward, Mars will always be visible and never set below the horizon.

> 3. Set the field of view to 5° to see a close-up of this configuration.
> 4. Return the field of view to 100°.

Mercury, Venus, Jupiter, and the crescent Moon are also visible in this view, as are the stars Castor and Pollux, the two brightest stars in the constellation Gemini, the Twins. For later reference, we can use the star Pollux to establish the position of Mars relative to the background stars.

5. Measure the angular separation between Mars and Pollux. (*Hint:* It might help to expand the field of view somewhat to separate Mars from the Sun in the sky.)

Question 1:   What is the angular separation of Mars from Pollux on August 10, 2002?

6. Run **Time Forward** for slightly more than one year, to **mid-October, 2003** (e.g., 10/8/2003), then click **Time Stop.** The exact date is not important.

TIP: The view is locked on to Mars. If you lose the lock, select **Synodic-A** from **Go/Observing Projects/Synodic Periods of Mars** and start time forward again.

Question 2:   What happens to the motion of Mars relative to the background stars on July 30, 2003, and September 28, 2003? You might want to run time backward and forward to watch this motion again.

This motion is discussed in more detail in the *Planetary Motion: The Retrograde Motion of Mars* project. As shown in that exercise, Mars reaches opposition on the date midway between the two dates mentioned in Question 2, i.e., on August 28, 2003.

The sidereal period of Mars is 687 days; so, starting from August 10, 2002, Mars completes one sidereal period on June 27, 2004. Thus, if we run time forward to June 27, 2004, we would expect Castor and Pollux to again be in the field of view because the sidereal period is defined as the orbital period relative to the background stars.

7. Run **Time Forward** to **June 27, 2004** (6/27/2004), and then click **Time Stop. Single Step Time Forward** or **Backward** to set the exact date, if necessary.
8. Measure the angular separation between Mars and Pollux.

If Pollux is not visible on your screen, check that the field of view is 100°.

Question 3:   What is the angular separation of Mars from Pollux on June 27, 2004? How does this value compare to the angular separation one sidereal period earlier, on August 10, 2002?

Question 4:   A comparison of your answers to Questions 1 and 3 shows that, in the view on the screen, Mars has *not* returned to the same position relative to the background stars after one sidereal period! Why is this?

Notice from your result in Questions 1 and 3 that there is nothing special about the position of Mars after one sidereal period, as seen from the Earth. Nothing about this position tells you that one sidereal period has elapsed. It is therefore not possible to measure the sidereal period of Mars directly by watching its motion relative to the stars or any other objects. The only period that we can measure directly from Earth is the **synodic period,** as shown below.

9. Run **Time Forward** to find the next conjunction of Mars with the Sun.
10. When Mars gets close to the Sun, change the field of view to 5° and the time step to 1 hour to determine the date of nearest approach precisely.

Question 5: On what date does this conjunction occur?

Question 6: How many days elapsed from the conjunction of August 10, 2002, to this conjunction?

B. Calculation of Synodic Period from Sidereal Period

As you saw above, observations of a planet can tell us the synodic period but not the sidereal period; therefore, historically, astronomers first measured the synodic period and then calculated the sidereal period from that. However, we will assume that we know the sidereal period and calculate the synodic period from it, to illustrate a very important point.

For superior planets, the equation relating the sidereal period, P, to the synodic period, S, and the orbital period of the Earth, E, is

$$\frac{1}{P} = \frac{1}{E} - \frac{1}{S}.$$

> This equation is derived and discussed in Box 4–1, Tools of the Astronomer's Trade, in Freedman and Kaufmann, *Universe*, 7th Ed., with a numerical example.

In this equation, P, E, and S must be in the same units. We can rewrite this equation by adding $\frac{1}{S}$ to both sides and subtracting $\frac{1}{P}$ from both sides to give

$$\frac{1}{S} = \frac{1}{E} - \frac{1}{P}.$$

Question 7: The sidereal period of Mars is $P = 687$ days, so using $E = 365.26$ days what value does the equation above give for the synodic period of Mars, in days?

Question 8: How does your measured time for this synodic period, from conjunction to conjunction, in Part A, compare to this calculated value for the synodic period?

You might notice in your answer to Question 8 that the calculated value and your measured value are somewhat different! This is not experimental error or scientific uncertainty; both numbers are in fact correct values. In Parts C and D below, you will see why these values are so different.

C. The View from Above the Solar System

We will now move our viewpoint to look down on the Sun and planets from above the solar system.

11. Select **Synodic-C** under **Go/Observing Projects/Synodic Period of Mars.**

The view shows Mars directly opposite the Sun from the Earth on August 10, 2002. Mars is therefore in conjunction with the Sun as seen from the Earth, as we noted in Part A.

Mars's orbit is more elliptical than that of the Earth, and as a result Mars is noticeably farther from the Sun and from the Earth's orbit in the position shown on the screen than it would be half a sidereal period later, when Mars would be diametrically across the Sun from its present position.

For future reference, note that there is a star just to the left of Mars in this view. If you place the cursor over this star, you should see that it is named Naos. We can use this star to show when Mars has completed one sidereal period. (Naos is, of course, a very distant background object. Its apparent nearness to Mars is an illusion produced by our viewpoint, looking down on the solar system.)

We will now watch Mars move through one synodic period as seen from the Earth, stopping the motion at a couple of places to compare this view with that in Part A.

12. Run **Time Forward** and watch the Earth "catch up" to Mars in its orbit. Click **Stop Time** when Mars is at opposition from the Earth, with the Earth on the Sun-Mars line. **Single Step Time Forward** or backward to find the exact date of opposition.

Question 9:   What is the date of opposition? Compare this value to the date quoted in Part A.

Question 10:   How far has the Earth moved around its orbit while Mars moves from conjunction to opposition? (To test your answer, you can click File/Revert to return to 8/10/2002, then set the time step to 1 year and single-step time forward by exactly this time interval. Reset the time step to one day and bring Mars back to opposition.)

13. Run **Time Forward** to June 27, 2004, and click **Stop Time**. **Single Step Time Forward** or **Backward** to set the date exactly.

You can see that Mars has now completed one sidereal period relative to the background stars because it is again close to the star Naos as seen from this fixed viewpoint above the solar system. The sidereal period is therefore a very natural period to use in describing the motion of planets around the Sun. Our only problem is that we have no direct way of measuring this period because we observe the planet from our moving Earth.

Notice in this view that the Earth and Mars are not opposite each other across the Sun, i.e., Mars is not yet in conjunction with the Sun as seen from the Earth. To complete one synodic period and bring Mars into conjunction with the Sun again, we need to run time forward again.

14. Run **Time Forward** to the date of conjunction found in Question 5, and click **Stop Time**. **Single Step Time Forward** or **Backward** to set the date exactly.

From this simulation you can see that the synodic period of Mars is longer than its sidereal period because Mars and the Earth are moving, and some extra time is needed after Mars completes one sidereal period to bring Mars and the Earth back into alignment across the Sun to complete one synodic period.

Question 11:   For the synodic period starting on August 10, 2002, how many days longer is Mars's synodic period than its sidereal period; i.e., starting from June 27, 2004, how many days elapse until the completion of one synodic period?

The simulation above shows why the synodic period of Mars is different from its sidereal period. We will now look at why your measured value of the synodic period in Question 6 differs from the standard value of 780 days.

D. The Synodic Period of Mars Beginning February 4, 2011

> 15. Select **Synodic-D** under **Go/Observing Projects/Synodic Period of Mars.**

On February 4, 2011, Mars and the Earth are again opposite each other across the Sun, so from the Earth we see Mars in conjunction with the Sun. This view, however, differs from that on August 10, 2002, in that Mars is near perihelion, i.e., near its closest point to the Sun and to the orbit of the Earth on this date, whereas on August 10, 2002, Mars was near aphelion (its farthest point from the Sun and the Earth's orbit).

One sidereal period (687 days) later, the date will be December 22, 2012.

> 16. Run **Time Forward** to December 22, 2012.

The view on the screen resembles that in Part B after one sidereal period (on June 27, 2004) in that Mars and the Earth are not opposite each other across the Sun, and some further time will have to elapse until they are. The view differs, however, in that Mars is now near perihelion instead of near aphelion.

> 17. Complete the synodic period by running **Time Forward** until Mars and the Earth are exactly opposite each other again; then click **Stop Time.**

The date should be April 17, 2013, although it may be difficult to determine this precisely using the view on the screen. You can check this date if you wish by selecting **Synodic-A** from **Go/Observing Projects/Synodic Period of Mars**, setting the date to April 17, 2013, and checking that Mars is indeed in conjunction with the Sun on this date. Then return to the view from above the solar system in the view **Synodic-D.**

Because Mars is close to perihelion, it is moving faster in its orbit than when it was near aphelion. The Earth will therefore take longer to "catch up" to the point directly opposite Mars than it did in Part B.

Question 12: How many days are there in one synodic period of Mars, for the synodic period starting February 4, 2011?

Question 13: How many days longer is the synodic period than the sidereal period of 687 days, for the synodic period starting February 4, 2011?

Question 14: How does your answer to Question 13 compare to your answer to Question 11 (more days, less days, or the same)?

The simulations in Parts B and C show that, when we measure the synodic period of Mars, our answer depends on the location of Mars in its orbit at the start of the synodic period. If Mars is close to aphelion, then the synodic period is less than the standard value, and if Mars is close to perihelion then the synodic period is greater than the standard value.

It would appear from this that the standard value must be a kind of average value. It is, in fact, the value we get by assuming **circular orbits,** and therefore **constant orbital speeds,** for the Earth and Mars. Different individual values for the synodic period will then be measured because of the elliptical orbits of the Earth and Mars, the specific value depending on where the Earth and Mars are in their orbits when we start the measurement. Since Mars has a much more elliptical orbit than the Earth, its position has the stronger influence on the measured value.

E. The Average Synodic Period of Mars

> 18. Open the file **Synodic-E** under **Go/Observing Projects/Synodic Period of Mars.**
> 19. Run **Time Forward** for eight conjunctions. This can be done by counting the number of times that the Sun passes Mars. The Sun moves in this view because you have the main view window locked on to Mars. (The eighth conjunction is in the year 2019.)
> 20. **Single Step Forward** or **Backward** to find the date of conjunction precisely.

> **TIP:** To accelerate time flow while time continues to run forward, click the time step interval in the control panel and use the + key to increment this value dynamically. When the Sun gets close to Mars in the year 2019, retard the rate of time flow to **1–day** intervals with the – key.

Question 15: What is the date of conjunction in 2019?

Question 16: How many days elapse from the conjunction in 2002 to that in 2019? (The conjunction in 2019 is later in the year than that in 2002, so you can find how many days apart these two dates are in a single year and add to that the number of days in 17 years. Rather than trying to account for individual leap years, it is accurate enough just to use an average value of 365.26 days per year.)

Question 17: What is the average synodic period of Mars over this time period? How does this value compare to the value you calculated in Question 6?

Your values in Questions 6 and 17 may not agree precisely because we are still measuring the value over a time period starting at a specific time, and the answer therefore still depends to some extent on the starting position of Mars and the Earth. To be absolutely precise, we would need to measure the period over many such cycles, or use a more complex mathematical procedure that accounts directly for the elliptical orbits of Mars and the Earth.

Nevertheless, averaging more than eight consecutive synodic periods smooths out the variations from one orbit to the next sufficiently that the two values should agree to within about a day.

F. Variation of Synodic Period with Distance from the Sun

In this section, you can investigate how the synodic period of a planet depends on its distance from the Sun. This can be done by using Kepler's third law and the planet's known distance from the Sun to find its sidereal period. You can then use the sidereal period to find the synodic period. We know that a planet's sidereal period increases with increasing distance from the Sun, from a short period (88 days) for Mercury, the closest planet to the Sun, to a very long period (248 years) for Pluto, the most distant planet from the Sun. The specific relationship is given by Kepler's third law,

$$P^2 = a^3 \qquad \text{Kepler's third law}$$

where P is the sidereal period in years and a is the planet's mean distance from the Sun in AU.

The equation in Part B,

$$\frac{1}{S} = \frac{1}{E} - \frac{1}{P} \, . \qquad \text{superior planets}$$

actually applies only to superior planets. As described in Box 4-1 of Freedman and Kaufmann, *Universe*, 7th Ed., the equivalent equation for inferior planets is

$$\frac{1}{P} = \frac{1}{E} + \frac{1}{S}.$$ inferior planets

or, subtracting $1/E$ from both sides of this equation,

$$\frac{1}{S} = \frac{1}{P} - \frac{1}{E}.$$ inferior planets

In each of these three equations, S (the synodic period), P (the sidereal period), and E (the Earth's period, 1 year or 365.26 days) must all be in the same units. Here, we will use years.

The Data Table below lists the six planets Mercury, Venus, Mars, Jupiter, Uranus, and Pluto and six hypothetical planets A, B, C, D, E, and F. The hypothetical planets are included to show how synodic period depends on distance from the Sun in regions of the solar system where there is no actual planet. Nevertheless, an object could orbit the Sun at these distances, such as an asteroid or a spacecraft.

For objects with orbital radii close to that of the Earth, to obtain meaningful sidereal and synodic periods we need to assume that the object is influenced only by the gravitational force of the Sun; i.e., the object does not feel any force from the Earth, even if it passes very close to the Earth. In reality this would not be true: the Earth's gravitational pull perturbs the orbit of any object passing close to it, sending the object into a slightly modified or even greatly modified orbit, depending on the distance of closest approach. This effect happens to asteroids and comets that pass close to any of the planets, particularly Jupiter, and has been used deliberately by NASA to save fuel by using a planetary slingshot effect to propel spacecraft to the outer planets. For the purpose of this exercise, however, we ignore this effect and look only at the effect of the Sun on the orbiting object.

The values of a, P, and S for the real planets can be found in the tables in the back of Freedman and Kaufmann, *Universe*, 7th Ed., and are filled in for you. For the hypothetical planets, the values of P and S need to be calculated from the specified distances from the Sun, using the method described above.

21. Use Kepler's third law to calculate the sidereal period for Planet A, and enter the value in the appropriate space in the Data Table.
22. Use the relationship between sidereal and synodic period to calculate the synodic period of Planet A, and enter the value in the appropriate space in the Data Table. Be careful to distinguish between inferior planets and superior planets.
23. Repeat the previous two steps for Planets B, C, D, E, and F.

DATA TABLE

Inferior planets	a (AU)	P (years)	S (years)
Planet A	0.10		
Mercury	0.39	0.2	0.32
Venus	0.72	0.62	1.6
Planet B	0.900		
Planet C	0.930		
Planet D	0.99990		

Superior planets	a (AU)	P (years)	S (years)
Planet E	1.0001		
Planet F	1.08		
Mars	1.52	1.9	2.1
Jupiter	5.20	12	1.09
Uranus	19.2	84	1.01
Pluto	39.5	248	1.00

A graph often represents the relationship between two variables with greater clarity than just a list of numbers. Use a sheet of finely divided graph paper to plot the synodic period, S, along the vertical axis against the distance from the Sun, a, along the horizontal axis. A convenient scale might be 1 year per large square for S and 1 AU per large square for a. Some data points will not fit on this graph, but the dependence of S on a should be clear from the data points that do fit.

Based on your graph and an examination of the data points in the Data Table that you did not plot on the graph, try answering the following questions.

Question 18: If we were to put a spacecraft into orbit around the Sun at a distance of 0.900 AU and then gradually increase the orbital distance closer and closer to 1 AU (say, 0.999 AU, 0.99999 AU, 0.999999999 AU, etc.), what would happen to the synodic period of the spacecraft? Why does this happen?

Question 19: Suppose we put the spacecraft into orbit around the Sun at a distance of 1.08 AU and then gradually decrease the orbital distance closer and closer to 1 AU (say, 1.001 AU, 1.00001 AU, etc.), what would happen to the synodic period of the spacecraft?

Question 20: Based on your answers to Questions 18 and 19, if someone were to ask, "What is the synodic period of an object orbiting the Sun at a distance of exactly 1 AU," what do you think the answer should be? (The mathematical process involved in obtaining this answer from the reasoning in Questions 18 and 19 is called "taking the limit of S as P approaches 1 AU"; that is, we are finding the limiting value of S as P becomes arbitrarily close to 1 AU.)

For the next two questions, think about the value of S shown in the Data Table for each planet compared to the value of P (the sidereal period of the planet) and the value of E (the sidereal period of the Earth, equal to 1 year).

Question 21: What happens to the value of S as the distance from the Sun, a, becomes very small? Why does this happen?

Question 22: What happens to the value of S as the distance from the Sun, a, becomes very large? Why does this happen?

G. Conclusions

In this project, you have seen how synodic and sidereal periods are defined, and why they differ. The sidereal period is measured by a hypothetical observer fixed in space, either on the Sun or suspended above the solar system, whereas the synodic period is measured by real observers (ourselves) watching the moving planet from a moving Earth.

The sidereal period can perhaps be visualized more easily than the synodic period, but only the synodic period can be measured directly.

You have also seen how to use an equation to calculate the sidereal period from the synodic period, although in this exercise you actually did the calculation in reverse, calculating the synodic period from the sidereal period.

The synodic period is not as easily measured as textbooks might lead you to believe, because the textbook derivation assumes circular orbits. If the orbits are elliptical (and all real orbits are), then the time from a given planetary configuration to the next occurrence of the same configuration (e.g., conjunction to conjunction) varies with time. You have seen why this happens, and have used a sound scientific method to obtain a more reliable result by measuring the average synodic period over a time of many years.

Finally, you have seen how a planet's synodic period depends on its distance from the Sun, and have looked at the "limiting values" of the synodic period for objects whose orbital radii are very small, very large, or are very close to 1 AU.

KEPLER'S THIRD LAW

14

Several thousand years ago, Greek philosophers followed logic and the principles of scientific investigation to develop models that described the motion of the "wandering stars" in the sky, the so-called planets. These geocentric models were based upon the understandable belief that the Earth was stationary. These models required many arbitrary assumptions and only over relatively short periods were they able to predict the future motion of the planets with reasonable accuracy. Over longer periods, the geocentric models required constant revision and became more and more complex.

In the 16th century, Copernicus was able to demonstrate that a model in which the planets orbited the Sun could explain these motions more simply. In his heliocentric model, he was able to show that the apparent complexity of the planetary motions arose from the fact that we were watching these planets from a moving Earth that also orbited the Sun.

By 1609, Johannes Kepler had developed this idea and had demonstrated that a simple model, in which the planets were required to obey only three laws, could describe the motion of the planets around the Sun:

1. The orbit of a planet about the Sun is an ellipse with the Sun at one focus.

2. The line joining a planet to the Sun sweeps out equal areas in equal intervals of time.

3. The square of the sidereal period, P, of a planet's orbit around the Sun is directly proportional to the cube of its semimajor axis, a; that is,

$$P^2 = \text{constant} \cdot a^3 \qquad \text{(Equation 1)}$$

where the constant is the same for all planets in the solar system.

For an elliptical orbit, the semimajor axis is half the longest diameter of the ellipse, that is, the distance from its center to one end. The sidereal period is the length of time the planet takes to complete one orbit, measured relative to the distant stars.

If we measure P in years and a in Astronomical Units (AU), then we can write Equation 1 more simply as

$$P^2 = a^3. \qquad \text{(Equation 2)}$$

A derivation of Equation 2 from Equation 1 is given in Appendix A near the end of this project.

Equation 1 is actually true for any object in orbit around any other object, providing that the mass of the orbiting object is much smaller than the mass of the object it is orbiting. Of course, the constant of proportionality will be different for different situations. For example, the satellites (moons) of all of the major planets obey Kepler's laws. In this project, after a brief glance at Jupiter itself, you can use *Starry Night*™ to investigate the motions of Jupiter's Galilean satellites. You can measure the semimajor axes and periods of each of these four moons and use these measurements to verify Kepler's third law. Then, with a little extra calculation, you can use your measurements to find the mass and mean density of Jupiter.

A. Jupiter and its Moons: An Initial Look

Before beginning the detailed measurement of the properties of its moons, it is worth taking the opportunity to explore several phenomena that occur during Jupiter's passage across our sky, such as the occultation of a star by the planet or the appearance of the shadow of a moon upon the planet's atmosphere. You can also examine the surface of this giant fluid planet briefly.

1. Launch *Starry Night*™.
2. Select **File/Preferences**. Open the **Cursor Tracking (HUD)** page in the Preferences dialog window and select **Name** and **Object type** from the **Show** list.
3. Select the view **Kepler-A** under **Go/Observing Projects/Kepler's Third Law**.

The field of view is 15 arc minutes wide (half the diameter of the Full Moon in our sky), and shows Jupiter with the four Galilean satellites near to it. The Hand Tool can be used to identify each moon in turn.

4. Select 30,000× as a time step interval and watch the Galilean moons orbit Jupiter.

> **TIP:** To return to the initial configuration, type
> **Ctrl + Z** (Cmd + Z on Macintosh) or select
> **Edit/Undo Time Step**.

Question 1: Which of the Galilean satellites moves in the smallest orbit and completes its orbit in the shortest time?

Question 2: Which of the Galilean satellites moves in the largest orbit and completes its orbit in the longest time?

The four satellites move around Jupiter in nearly circular orbits in the plane of Jupiter's equator, and because Jupiter's equatorial plane almost coincides with the ecliptic (that is, with the Earth's orbital plane), we see these orbits almost edge-on. Thus, we see each satellite apparently moving back and forth past Jupiter in almost a straight line, the nearest ones to the planet moving fastest, as expected from Kepler's laws. At this time and with this geometry, three of the four moons undergo regular **occultations,** in which the moon passes behind the planet. Disappearance is known as **ingress** while reappearance is known as **egress.** These three moons also **transit,** or pass in front of, the planet.

Several features of these motions are worthy of note. After two of these moons are occulted by Jupiter, they suddenly reappear some distance to the east of Jupiter, while after a third moon is occulted it reappears at Jupiter's east limb but then appears to wink out for a short period. This effect can be examined in more detail by zooming in on Jupiter.

Question 3: In view of the geometry of Jupiter and its moons, why do these moons appear somewhat dimmer at certain times following egress from occultation?

5. Select **Kepler-A2** under **Go/Observing Projects/Kepler's Third Law**.

Ganymede, the third most distant of the Galilean moons from Jupiter, can be seen just before ingress.

6. Run **Time Forward** and observe the unfolding sequence of events.

Watch Ganymede particularly as it moves ahead in its orbit, after it has been occulted by and has reappeared from behind Jupiter.

Europa, the second of the Galilean moons from Jupiter, appears from the left and transits the planet, starting at about 7:35 P.M. on February 23, 2001.

Io, the innermost Galilean moon, appears on the right of the screen and goes into occultation at about 11:40:00 P.M. Watch this moon particularly as it reappears from behind the planet.

7. Select **Kepler-A3** under **Go/Observing Projects/Kepler's Third Law.**

In this view, Europa is part way through a transit, passing in front of the structured atmosphere of Jupiter.

8. Run **Time Forward** to watch the unfolding sequence of events.

Europa soon completes its transit, and leaves the west limb of Jupiter. As this happens, you will see the shadow of this moon begin to traverse Jupiter's surface. The positions of Europa and its shadow show the direction of the Sun at this time.

One interesting puzzle about shadows of this kind is that, in infrared light, this shaded area appears hotter than the surrounding atmosphere. The reason for this is probably the momentary blocking of sunlight that causes clouds in Jupiter's atmosphere to dissipate, allowing infrared detectors to see more deeply into warmer regions of Jupiter's atmosphere. A related effect is the observation that the relatively cloud-free regions known as "belts" around Jupiter appear dark in visible images but bright in infrared images, suggesting that they are warmer than the adjacent optically bright (but infrared-dark) zones. You can search for the shadows of other moons, particularly Io and Ganymede, as they transit the planet.

At this magnification, the patterns and spots on the planet's surface are clearly evident, including the Great Red Spot in Jupiter's southern hemisphere. In reality, this image would show differential rotation, with the equator rotating faster than the poles, and these patterns would be changing and evolving with time because of this motion. Nevertheless, this simulation shows the stunning patterns and colors on Jupiter's surface.

There is one more event that you can witness with this simulation, namely the occultation of a star by Jupiter. Observations of the star and its spectrum during such an event, particularly as Jupiter's atmospheric layers occult the star, have been used to determine the chemical makeup and structure of the planet's atmosphere.

9. Select **Kepler-A4** under **Go/Observing Projects/Kepler's Third Law.**
10. Run **Time Forward** and watch both Io and its shadow traverse the planet while Jupiter passes in front of the star, TYC1258-377-1.

By coincidence, at the beginning of this simulation, Io's orbital motion toward the west almost exactly compensates for Jupiter's orbital motion toward the east, so that Io appears to be stationary in our sky. If you continue to watch, Io will gradually start to move eastward as it approaches the western "end" of its orbit as viewed from the Earth. In this simulation, the star image appears larger than it would appear in the real sky.

Thus, you can see that there are always lots of interesting events and activity in the vicinity of Jupiter as this planet and its moons continue their ceaseless journeys through space.

B. Motions of the Galilean Satellites of Jupiter

You can verify Kepler's third law, the relationship between semimajor axis and orbital period, using further observations of the detailed motions of the Galilean moons of Jupiter.

The orbits of the four Galilean moons are almost precisely circular (with eccentricity 0.01 or less), so the length of the semimajor axis is simply the radius of the orbit (half the diameter). To measure the length of the semimajor axis of any moon's orbit, therefore, we can measure the length of the line from the center of Jupiter to the moon when this moon is at its farthest position from Jupiter, as seen on the screen.

11. For each of the four Galilean moons, use the procedure outlined in the following steps to determine the semimajor axis of its orbit.
12. Select **Kepler-B** under **Go/Observing Projects/Kepler's Third Law.**
13. Open the contextual menu over the specific moon and select **Orbit.**
14. If necessary, adjust the zoom factor of the view so that the moon's orbit is visible in the view.
15. Manipulate the time controls to place the selected moon at one extreme of its orbit.
16. Measure the angular spacing between the center of Jupiter and the selected moon. Record this distance in Data Table 1 below.
17. Convert this angle into arc seconds (1' = 60") and record this number in Data Table 1 below.

DATA TABLE 1: Semimajor Axis

Moon	Semimajor axis, a (' ")	Semimajor axis, a (")
Io		
Europa		
Ganymede		
Callisto		

We can measure the orbital period by measuring the time the satellite takes to move from a reference position on its orbit and back to the same position. For the chosen observing time in March 2001, an excellent reference for the three inner moons is the time when the moon appears to cross the limb or edge of Jupiter. Callisto, the outermost Galilean moon, does not cross Jupiter's limb in March 2001 because of the relative geometry of Earth and Jupiter, but we can record the times when it passes over the spin axis of Jupiter.

18. Select **Kepler-B2** under **Go/Observing Projects/Kepler's Third Law.**

The view shows Io on the west limb of Jupiter, about to begin ingress into occultation.

19. Record the date and time under Date 1 and Time 1, respectively, for Io in Data Table 2 below.

The view is locked onto Io. Thus, Jupiter will appear to move when time advances. Io will pass through occultation, return and transit Jupiter, and then approach occultation again to complete one orbit.

20. **Run Time Forward.**
21. Stop time when Io once again approaches the west limb of Jupiter, about to begin ingress into occultation.
22. Change the time step to **1 minute**, and **Single Step** forward or back to find the exact time when the satellite contacts the limb of Jupiter.
23. Record the date and time under Date 2 and Time 2, respectively, for Io in Data Table 2.
24. Select **Kepler-B3** and repeat these measurements for Europa.
25. Select **Kepler-B4** and repeat these measurements for Ganymede.
26. Select **Kepler-B5**. Jupiter's rotation axis is shown, and Callisto appears exactly over the south pole of the planet.
27. Repeat the above measurements for Callisto, using Jupiter's south polar axis as the reference point.
28. Using the method described below, calculate the orbital period of each moon in days and hours, and then convert the period to hours. Enter your results into Data Table 2.

Calculation of Orbital Period from Limb-Crossing Times

To determine the orbital period of each moon, you need to calculate the difference in time between the first and second limb passages. This has to be done carefully. For this purpose, it is perhaps helpful to draw a diagram to represent the time interval, such as the one shown below. The horizontal direction represents time, with vertical marks showing midnight for each day. The spacings between these lines will then represent days and you can number them by date, between Date 1 and Date 2 from the Data Table above. You can now mark Times 1 and 2 in the appropriate days on your diagram. It is now easy to envision the time and determine the interval between these two times. You need to add the remainder of the first day, the number of full days that follow and the time in the last day. The diagram shows this pattern for a fictitious data set, where the start time is 12:30:00 P.M. on March 3 and the end time is 05:14:00 A.M. on March 7.

March 3	March 4	March 5	March 6	March 7
	← Date and Time 1		Date and Time 2 →	

In this example, you can determine the time as follows.

March 3	Remaining time	= 24:00–12:30	= 11 hours, 30 minutes
March 4–6	3 full days	= 3 × 24 hours	= 72 hours,
March 7	Time	= 5 hours, 14 minutes	= 5 hours, 14 minutes

Total time = 88 hours, 44 minutes

To convert this time to hours, divide the minutes (44 in the example above) by 60 to obtain a fraction of an hour (0.73), and add this to the number of hours (88.73 hours).

DATA TABLE 2: Limb Crossing Times and Calculated Orbital Period

	Date 1 (m/d)	Time 1 (h/m/s)	Date 2 (m/d)	Time 2 (h/m/s)	P (hours/minutes)	P (hours)
Io						
Europa						
Ganymede						
Callisto						

Before using your data to verify Kepler's third law, the following questions outline an interesting relationship between the periods of these moons.

Question 4: What number do you get if you divide the orbital period of Europa by the orbital period of Io? Is this number close to an integer? Which integer?

Question 5: What number do you get if you divide the orbital period of Ganymede by the orbital period of Io? Is this number close to an integer? Which integer?

Question 6: What number do you get if you divide the orbital period of Ganymede by the orbital period of Europa? Is this number close to an integer? Which integer?

Question 7: From your answers to the above questions, is there any clear synchronicity between the orbital periods of Io, Europa, and Ganymede?

C. Verifying Kepler's Third Law

Equation 1 represents Kepler's third law for objects orbiting a much more massive object. If the orbits are circular, then P in Equation 1 is the orbital period and a is the orbital radius. If we write the constant in Equation 1 as k, then Kepler's third law becomes

$$P^2 = k \cdot a^3$$

where the numerical value of k depends on the units that are used for P and a.

Because

$$k = \frac{P^2}{a^3}$$

we can verify Kepler's third law by demonstrating that the value of k is the same for all of the Galilean moons. Data Table 3 below allows you to calculate the value of k for each of these moons.

29. Copy the values of P and a from Data Table 2 over to Data Table 3 below.
30. Calculate P^2, a^3, and the value of k for each moon.

DATA TABLE 3: Verification of Kepler's Third Law

Moon	P	P^2	a	a^3	$k = \frac{P^2}{a^3}$
Io					
Europa					
Ganymede					
Callisto					

If, from your data, k has the same value for all four moons, then you have verified Kepler's law as it is applied to the Galilean moons of Jupiter. For the data used here, the units of k are $\dfrac{(\text{hours})^2}{(\text{arc seconds})^3}$.

      Question 8:   Do the moons of Jupiter obey Kepler's law, according to your data?

D. Measuring the Mass of Jupiter

> **Note:** Parts D and E are more mathematical than parts A to C above.

For the special case of an object moving in a circular orbit around an object whose mass is much larger than itself, Kepler's third law becomes

$$R^3 = \left[\frac{GM}{4\pi^2} \right] \cdot P^2 \qquad \text{(Equation 3)}$$

where R is the radius of the satellite's orbit, P is the satellite's orbital period, M is the mass of the central object (in the present case, Jupiter), and $G = 6.68 \times 10^{-11}$ N m^2/kg^2 is the gravitational constant. (If you are interested in how this equation arises, a derivation is given in Appendix B at the end of this project.) Because the Galilean moons are in circular orbits and the mass of each Galilean moon is less than 1/1000 of the mass of Jupiter, the required conditions are fulfilled for these objects.

    Multiplying both sides of Equation 3 by $4\pi^2$ and dividing both sides by GP^2, then switching sides, gives

$$M = \frac{4\pi^2 R^3}{GP^2} \qquad \text{(Equation 4)}$$

In this equation, if we can calculate R and P in units of meters and seconds, respectively, from our measurements of the moon's motions, we can derive the mass of Jupiter. G is a universal constant with a value of 6.68×10^{-11} N m^2/kg^2. The mass of Jupiter M is then derived in kilograms.

    You can use any of the satellites to calculate Jupiter's mass, but it is probably best to choose Callisto. This is because Callisto has the largest values of R and P, so the uncertainties in measurement are the smallest, relative to R and P themselves.

31. Transfer Callisto's period, P, in hours, from Data Table 2 to Data Table 4 below.
32. Multiply this period by 3600 (the number of seconds in one hour) to find the period in seconds. Enter the result into Data Table 4 below.

33. Set the date to near the midpoint between the start and end of the time when you measured Callisto's period.
34. Open the Info pane for **Jupiter**. Under the **Position in Space** layer, find the value given for **Distance from Observer** and record this in Data Table 4 as the Distance to Jupiter (in AU).
35. Multiply this distance by 1.49×10^{11} (the number of meters in 1 AU) to find the distance to Jupiter (in meters), and enter the result into Data Table 4.
36. Transfer the semimajor axis of Callisto's orbit, a, in arc seconds, from Data Table 1 to Data Table 4.
37. Divide this angle by 206,265 (the number of arc seconds in one radian) to find the semimajor axis in radians, and enter the result into Data Table 4. We will denote this angle (measured in radians) by the symbol α.
38. Use the small-angle formula (Equation 5 below) to calculate Callisto's semimajor axis in meters, R, and enter the result into Data Table 4.
39. Finally, substitute R, P, and G into Equation 4 to find the mass, M_J, of Jupiter, and enter the result into Data Table 4.

> **TIP:** An easy way to open the Info pane for Jupiter is to open the contextual menu over the planet and select **Show Info.** Alternatively, double-click over the planet.

To find R in meters, use the small-angle formula,

$$R = \alpha d \qquad \text{(Equation 5)}$$

where α is the angle taken up by the semimajor axis in radians, and d is the distance to Jupiter in meters.

DATA TABLE 4: Calculation of Jupiter's Mass from Callisto's Motion

Period in hours (from Table 2)	
Period in seconds	$P =$
Distance to Jupiter (in AU)	
Distance to Jupiter (in meters) (1 AU = 1.49×10^{11}m)	$d =$
Semimajor axis of Callisto's orbit (in arc seconds) from Table 1	
Semimajor axis (in radians) (1 radian = 206,265 arc seconds)	$\alpha =$
Semimajor axis (in meters): $R = \alpha d$	$R =$
Mass of Jupiter	$M_J =$

Question 9: How does your result compare to Jupiter's accepted mass of 1.9×10^{27} kg?

E. Measuring the Mean Density of Jupiter

The mean density, ρ, of a planet is equal to the mass of the planet divided by its volume:

$$\rho = \frac{M}{V} = \frac{M}{\left(\frac{4\pi R^3}{3}\right)} \qquad \text{(Equation 6)}$$

where M and R are the planet's mass and radius (half its diameter), respectively, and $V = \frac{4\pi R^3}{3}$ is the volume of a sphere. Here we are assuming that Jupiter is close enough in shape to a sphere that we can neglect its rotational flattening.

Because we have already measured Jupiter's mass, we now need only to measure its radius in order to calculate the mean density of Jupiter.

> If you did not obtain a mass in Part D, you can use the accepted value given in Question 9.

40. Select **Kepler-E** under **Go/Observing Projects/Kepler's Third Law.**
41. Use the **Hand Tool** to measure the angular radius of Jupiter (in seconds of arc) on this date, and enter the result into Data Table 5 below.
42. Divide this angle by 206,265 (the number of arc seconds in one radian) to find the angular radius of Jupiter (in radians), and enter the result into Data Table 5. We will denote this angle (in radians) by the symbol α_J.
43. Open the **Info** tab for Jupiter. Under the **Position in Space** layer, find the value given for **Distance from Observer** and record this in Data Table 5 as the Distance to Jupiter (in AU).
44. Multiply this distance by 1.49×10^{11} (the number of meters in 1 AU) to find the distance to Jupiter (in meters), and enter the result into Data Table 5.
45. Use the small-angle formula (Equation 5) above to find the radius of Jupiter (in meters) by multiplying α_J in radians by d in meters, and enter the result into Data Table 5.
46. Now substitute M_J in kilograms from Table 4 and your calculated value of R in meters into Equation 6 to find the mean density, ρ, of Jupiter.

DATA TABLE 5: Calculation of the Mean Density of Jupiter

Mass of Jupiter (from Table 4)	$M_J =$
Angular radius of Jupiter (in seconds of arc)	
Angular radius of Jupiter (in radians)	$\alpha_J =$
Distance to Jupiter (in AU)	
Distance to Jupiter (in meters)	$d =$
Radius of Jupiter (in meters): $R_J = \alpha_J \cdot d$	$R_J =$
Mean density of Jupiter	$\rho =$

Question 10: How does your result compare to Jupiter's accepted mean density of 1300 kg/m³?

Question 11: How does your result compare to the density of water (1000 kg/m³), the density of rock (about 2800 kg/m³), and the mean density of the Earth (5500 kg/m³)?

Question 12: What does your answer to Question 11 suggest about the bulk composition of Jupiter?

Appendix A

The following steps show a derivation of Equation 2 from Equation 1 in the Introduction.

Applying Equation 1 to the Earth gives

$$P^2_{\text{Earth}} = k \cdot a^3_{\text{Earth}} \qquad \text{(Equation 7)}$$

where k is a constant.

Then, if we divide both sides of Equation 1 by the square of the Earth's sidereal period and use Equation 2, we get

$$\frac{P^2}{P^2_{\text{Earth}}} = \frac{k \cdot a^3}{P^2_{\text{Earth}}} = \frac{k \cdot a^3}{k \cdot a^3_{\text{Earth}}} \qquad \text{(Equation 8)}$$

The constant cancels in the last term of Equation (8). Then using $\dfrac{a^n}{b^n} = \left(\dfrac{a}{b}\right)^n$, we have

$$\left(\frac{P}{P_{\text{Earth}}}\right)^2 = \left(\frac{a}{a_{\text{Earth}}}\right)^3 \qquad \text{(Equation 9)}$$

Finally, if we measure P in years and a in AU, then $P_{\text{Earth}} = 1$ and $a_{\text{Earth}} = 1$, and Equation 9 becomes simply

$$P^2 = a^3. \qquad \text{(Equation 2)}$$

Appendix B

The following steps show a modern derivation of Kepler's third law, for the special case of an object in a circular orbit around another object of much higher mass. (That is, a situation in which the mass of the orbiting object can be neglected compared to the mass of the object it is orbiting.)

An object's velocity is determined by the object's speed and direction of travel. Acceleration is the rate of change of velocity. Therefore, acceleration can involve a change in speed, a change in direction, or both.

An object moving in a circular orbit has a speed that is constant. Nevertheless, the object is accelerating because its direction of travel is continuously changing. This centripetal acceleration due to circular motion is given by

$$a_{\text{cent}} = \frac{v^2}{R} \qquad \text{(Equation 10)}$$

where v is the object's orbital speed and R is the radius of the orbit.

The object remains in orbit because there is one (and only one) force acting upon it, the force of gravity. To see that only one force is acting on it, consider Newton's first law: *"Any object remains at rest or moves in a straight line at constant speed if no force is acting on it."* Thus, if no force acted on an object in orbit, or if there were inward and outward forces that "cancelled," then the object would move in a straight line, NOT in a circle. Motion in a circle requires a force. This force is supplied by gravity.

Newton's law of gravitation says that the gravitational force between two objects of masses M and m, a distance R apart, is given by

$$F_{\text{grav}} = \frac{GMm}{R^2}, \qquad \text{(Equation 11)}$$

where $G = 6.68 \times 10^{-11}$ N m²/kg² is the gravitational constant. We will take m to be the mass of the orbiting object, and M to be the mass of the object that it is orbiting.

Finally, Newton's second law states that the sum of all forces acting on any object is equal to the object's mass times its acceleration. Here, there is only one force acting, the gravitational force, so we simply have

$$F = ma \qquad \text{(Equation 12)}$$

Substituting $a = a_{\text{cent}}$ and $F = F_{\text{grav}}$ from Equations 10 and 11 into Equation 12 then gives

$$\frac{GMm}{R^2} = m\frac{v^2}{R}. \qquad \text{(Equation 13)}$$

In practice, the two R's in Equation 13 are not the same because the masses M and m actually orbit around a common center of mass, called the barycenter. (For example, in the Earth-Moon system the center of mass is about 5000 km from the center of the Earth [or about 1000 km below the surface of the Earth], along the line joining the centers of the Earth and the Moon.)

However, if m is "small" compared to M, as is the case with the Galilean satellites orbiting Jupiter, then the displacement of the center of mass from the center of M is also "small" compared to the distance between m and M. In this case, we can regard the two R's in Equation 13 as being essentially equal. Multiplying both sides of Equation 13 by R then gives

$$\frac{GMm}{R} = mv^2 \qquad \text{(Equation 14)}$$

The orbital speed of the object can be found by dividing the distance around the orbit by the time taken to complete the orbit. The distance around the orbit is the circumference, given by $C = \pi D = 2\pi R$, where D is the orbital diameter and R is the orbital radius. The time taken to complete the orbit is the orbital period, P. Then

$$v = \frac{C}{P} = \frac{2\pi R}{P} \qquad \text{(Equation 15)}$$

Substituting Equation 15 into Equation 14 gives

$$\frac{GMm}{R} = m\left(\frac{2\pi R}{P}\right)^2$$

The mass of the orbiting object, m, cancels. Then, squaring the quantity in parentheses gives

$$\frac{GM}{R} = \frac{4\pi^2 R^2}{P^2}$$

Finally, multiplying both sides by P^2R and dividing both sides by GM gives

$$P^2 = \left(\frac{4\pi^2}{GM}\right)R^3 \qquad \text{(Equation 3)}$$

Equation 3 has the same form as Equation 1, which means that it is a statement of Kepler's third law. The derivation shows that this particular form applies to the special case of a circular orbit, where the mass of the orbiting object is small compared to the mass of the objected around which it is orbiting.

COMETS AND THEIR ORBITS 15

Comets are perhaps the most enigmatic of objects in our solar system. Their appearance in the sky has invoked fear as well as admiration in many civilizations. The tail of a bright comet can stretch across half the visible sky and be as bright as most stars, if only for a few days. Nevertheless, there is little substance to these ephemeral visitors.

Comets are either in long, stable elliptical orbits around the Sun or are first-time visitors to our solar system, making a single looping pass around the Sun. A typical comet consists of

> Comets are discussed in Sections 17–7 and 17–8 of Freedman and Kaufmann, *Universe*, 7th Ed.

a rather small ball of ice, dust, and rock. As this insignificant snowball comes close to the Sun, it is partially melted and evaporated by the Sun's radiation to release atoms, molecules and dust grains into a large cloud around its nucleus. These components either reflect sunlight or emit intrinsic light when excited by solar ultraviolet radiation. Much of this thin cloud of released material is blown outward from the Sun into a long tail in the anti-sun direction by the action of the solar wind and sunlight. The detailed shape and structure of the tail reflects the action of this solar wind upon the ionized atoms and molecules and of radiation pressure of sunlight upon the larger grains of dust. Thus, even with this very small inherent mass, a comet can stretch itself across a significant region of our sky and become disproportionately bright during its brief visit to the Sun's vicinity.

In this project, we will watch the development and alignment of the celebrated and much-studied **Halley's comet**, named for Edmund Halley in recognition of his prediction of its return in 1758. This prediction was shown to be correct when the comet was "discovered" on Christmas night of 1758, though regrettably, this observation occurred long after Halley's death. Halley's insight that the appearance of a major comet in 1531, 1607, and 1682 was in fact a returning object and the detailed calculation of its orbit using Newton's newly developed theory of gravitation was a major milestone in the development of a rational explanation for the behavior of objects in the solar system. Halley's comet has been observed in every one of its returns to the vicinity of the Sun on a 76-year cycle since 239 B.C.

The simulations in this project will show the general appearance and behavior of

> Tail structure is shown in Figs. 17–18, 17–23, and 17–24 of Freedman and Kaufmann, *Universe*, 7th Ed.

Halley's comet. You will observe the growth, decay and direction of its tail during the comet's last appearance in 1986, although the detailed and changing structure of this tail under the influence of the variable solar wind is difficult to reproduce in simulation.

> Kepler's second law is discussed in Section 4–4 of Freedman and Kaufmann, *Universe*, 7th Ed.

You will also be able to verify Kepler's second law of planetary motion with this comet. This law, originally stated by Kepler in the form: *"As an object orbits the Sun, the line joining planet to Sun will sweep out equal areas in equal times,"* leads to the conclusion that the speed of any object moving under the influence of the Sun will be higher, the closer the object is to the Sun. Halley's comet moves in a long elliptical orbit and its distance from the Sun, and therefore its orbital speed, changes over a wide range. The motion of this spectacular object thus provides a good general test of this law.

A. Observation of Halley's Comet from Earth

We will first observe Halley's comet from the Earth's surface. The chosen site is Caracas, Venezuela, at a latitude of about 10° N. This simulation shows a predawn view of the southeast sky in March 1986 during the close approach of the comet to the Sun.

1. Launch *Starry Night*™ and select **Comet-A** from **Go/Observing Projects/Comets**.
2. Set the **Cursor Tracking (HUD)** preferences under **File/Preferences** so that the HUD displays the **Name** and **Object type**.

The view is toward the SE from Caracas, across a flat horizon placed low down in a wide field of view, at 3:00:00 A.M. on March 7, 1986. The time step is set to 1 minute. At this time, the Milky Way is spread across the sky. Halley's comet can be seen just above the horizon between S and SE, with its tail extended upward, away from the direction of the rising Sun.

3. Run **Time Forward** and stop time at about 06:30:00 A.M., just after sunrise.

TIP: An alternative technique for watching this time sequence is to click the minutes field of the Time in the control panel and hold down the + key on the keyboard to advance time in 1-minute intervals.

As time advances, you will see Halley's comet rise majestically from the horizon, along with the star background. As sunrise approaches, the comet's tail slowly fades into the morning twilight. Scattering of sunlight from small dust particles produces the blue color of its tail. We infer from the color of this scattered light that the typical size of these dust particles is equivalent to the smoke particles from a campfire.

4. Adjust the time to **5:42:00 A.M.**
5. Use the angular measurement feature of the **Hand Tool** to draw a line from the Sun through Halley's comet. (If the sky is too bright for the comet to be visible, select **View/Hide Daylight**.) Note the alignment of the comet's tail in the anti-sun direction.

Sidereal time is discussed in Box 2–2, Tools of the Astronomer's Trade in Freedman and Kaufmann, *Universe*, 7th Ed.

It is possible to watch the movement of the comet over a longer period against the background stars by stepping time in intervals of **sidereal** days. This will keep the stars in exactly the same position in the sky, night-by-night. We can start this sequence at 3 A.M. on March 7, 1986, when the Sun is below the Eastern horizon and the comet is moving away from the Sun after its close approach in late February.

6. Select **Comet-A2** under **Go/Observing Projects/Comets**. This will reset the date to March 7, 1986, the time to 03:00:00 A.M., the time step interval to 1 sidereal day, and place the comet in the lower left corner of the view.
7. Use the **Single Step Forward** to advance the time in 1-sidereal day steps to April 13, 1986.

In this sequence, it is obvious that the direction of the comet's tail is not controlled by motion of the comet itself because the tail precedes the comet nucleus on this section of its orbit. In fact, the tail is controlled by the direction of sunlight and the prevailing solar wind.

You can watch the majestic sweep of the comet as it moves away from the Sun after this close approach.

8. Adjust the view so that Halley's comet is near the left edge of the screen on April 13, 1986.
9. Use the **Single Step Forward** button to advance time in 1-sidereal day steps until you can no longer see the comet's tail.

Question 1: Approximately when does Halley's comet's tail disappear?

As the simulation progresses in steps of 1 sidereal day, each day's observation occurs about 4 minutes earlier than that of the previous night until eventually, daylight interferes. If we remove the blue sky of daylight, we can observe both the comet's approach to the Sun and its recession again into space.

10. Select **Comet-A3** under **Go/Observing Projects/Comets.**

The view is of the SE sky from Caracas at 8:30 A.M. on January 7, 1986, with daylight off. At this time, the Sun is about 35° above the horizon, along with the planets Venus and Mercury. The planet Jupiter can also be seen just above the eastern horizon, close to the comet.

11. Identify the Sun, Venus, Mercury, and Jupiter with the **Hand Tool.**
12. **Step Time Forward** in steps of 1 sidereal day.

As time progresses in time steps of 1 sidereal day, the Sun moves eastward toward its sunrise position, accompanied by its entourage of planets. Meanwhile, the comet rises as it approaches the Sun and we see the foreshortened tail pointing away from the Sun at the time of the comet's closest approach to the Sun in February.

By the end of March, the comet has moved to its position of closest approach to Earth and shows us a spectacular side-on view. After this, the tail becomes progressively smaller. While part of this apparent reduction is foreshortening as we see the tail from an end-on view, there is also a real reduction in the length of the tail as the comet moves away from the Sun. Solar heat falling upon the comet's nucleus is reduced as it moves away from the Sun and less of this nucleus is being evaporated to form the tail.

13. When the date reaches about **April 8, 1986,** move the view direction to SW.
14. Resume stepping forward at 1-sidereal day intervals to watch the comet move away into space. Stop time advance on **August 1, 1986,** when the comet is almost invisible.

After this brief visit to our vicinity, Halley's comet will return again only after about another 75 years. Most of this time will be spent as a cold, icy nucleus, unaccompanied by a tail and invisible from Earth.

This sequence has given you a panoramic view of the full sweep of the comet and its tail during the close approach to the Sun, projected against a fixed sky without the hindrance of daylight, an opportunity not possible in real life!

The tail is at its most spectacular in late March and early April as the comet comes closest to Earth. As we shall see from other viewpoints, this is not the time of maximum intrinsic tail length. This will be when the comet is closest to the energy source that melts the ice to produce its tail, our

Sun. It is also interesting to remember that the tail changes materially day-by-day such that, were you to be watching the comet in real life, you would be viewing a new tail every day! For all its glorious appearance, the amount of material within this tail is miniscule and could probably be packed into a small suitcase. Every atom, molecule, or grain of dust is emitting light or scattering sunlight toward us to produce this great sight.

You can repeat this sequence and use the **Info** facility to monitor the distance of Halley's comet from the Earth and the Sun to determine the dates of closest approach to each of these objects.

15. Select **Comet-A3** under **Go/Observing Projects/Comets** to return to the initial conditions of this sequence.
16. Open the contextual menu for Halley's comet and select **Show Info.**
17. In the **Info** pane, expand the **Position in Space** layer to display distances of Halley's comet from the observer and from the Sun.
18. Advance time in steps of 1 sidereal day and note the distance of the comet from the Sun and from the Earth shown in the **Position in Space** layer of the Info pane. Use your observations to answer the following questions.

Question 2:   What is the date of closest approach of the comet to the Sun?

Question 3:   How far is the comet from the Sun at closest approach?

Question 4:   What is the date of closest approach of the comet to the Earth?

Question 5:   How far is the comet from Earth at closest approach?

B. Further Observations of Halley's Comet

The next sequences allow you to observe Halley's comet from novel perspectives.

19. Close the **Info** pane.
20. Select **Comet-B** under **Go/Observing Projects/Comets.**

In this view, you are hovering about 2.1 AU above the Sun's north pole on December 20, 1985, at 12:00 UT. The planets Mercury, Venus, and Earth are shown in the correct positions on their respective orbits. The Earth is to the right of the screen and Halley's comet is in the lower right of the view. Note that Halley's comet lies beyond the Earth's orbital distance, with its tail aligned in the anti-sun direction as expected. As time advances in this mode, the objects in the solar system move while the background stars remain stationary.

21. Run **Time Forward** to watch the comet progress across the inner planetary system.

Running time forward with the time interval of 6 hours will show the comet's tail grow and shrink again under the varying influence of the Sun's heat as the comet traverses the inner solar system. You can see that this tail maintains its anti-sun alignment as it grows and decays.

22. Select **Comet-B** under **Go/Observing Projects/Comets.**
23. Manipulate time forward and backward with the time controls to answer the following questions. Use single steps forward and backward in time as necessary to refine your answers.

Question 6: On what date does the comet appear to cross the Earth's orbit on its way into the inner solar system?

Question 7: On what date does the comet appear to cross the Earth's orbit as it leaves the inner solar system?

Question 8: Is Halley's comet orbiting the Sun in the same or in the opposite direction to the planets?

Question 9: In which part of the Comet's orbit does it come closest to the Earth, incoming or outgoing?

24. Select Comet-B2 from Go/Observing Projects/Comets.

The view is looking edge-on along the plane of Earth's orbit at the inner solar system from a position in space about 2.1 AU from the Sun. The Earth can be seen to the right of the screen. Meanwhile, the comet is at the right edge of the view, somewhat outside the Earth's distance from the Sun at the start date of December 20, 1985.

On this date, the comet has already moved through the ecliptic plane as it approaches the Sun and is above the north side of this plane.

25. Run Time Forward to watch the comet's progress from this vantage point.

You will see that the comet remains north of the ecliptic plane during this passage inside the Earth's orbit, passing south again only late in this encounter with the Sun. Again, its tail maintains the anti-sun alignment.

Question 10: When does the comet pass southward through the ecliptic plane?

We can now take one last look at the passage of Halley's comet across the inner solar system from an oblique angle to the ecliptic plane to show its three-dimensional path.

26. Select Comet-B3 from Go/Observing Projects/Comets.

The view is from a point above the plane of the solar system about 2 AU from the Sun on December 1, 1985, when the comet is still beyond the orbit of Mars. We can watch the comet progress from this point toward the Sun, sweeping majestically through the inner solar system on this hasty visit from distant space.

27. Run Time Forward until about 5/03/1986 A.D., when the comet is well beyond the orbit of Earth.

C. Kepler's Second Law and Cometary Motion

Halley's comet moves through a wide range of distances from the Sun as it traverses its orbit. Thus, we can use measurements of its motion to verify Kepler's second law over this range of distances. As stated above, Kepler defined this law in the following terms. *"A line joining a planet and the Sun sweeps out equal areas in equal intervals of time."* Another way of expressing this law is to say that the instantaneous speed of the comet is inversely proportional to its distance from the Sun. We can verify this law by observing the comet from the unique viewpoint of the Sun.

The relationship between the area swept out by the comet and measurable quantities can be derived easily. The relevant observations are the **angular speed** of the comet against the background stars, and the **distance** of the comet from the Sun.

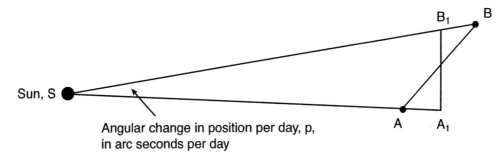

Let us assume that we make measurements over intervals of 1 day. In the diagram above, the comet moves along its orbit from A to B in this time, changing its position against the background stars as seen from the Sun by the angle p. The area swept out by the Sun-comet line over this time is the tri angle SAB. Because the angle p is very small, we can approximate this

> Small-angle geometry is discussed in Box 1–1, Tools of the Astronomer's Trade, in Freedman and Kaufmann, *Universe*, 7th Ed.

area with that of triangle SA_1B_1. The area of triangle SA_1B_1 is equal to one-half of the product of its base (A_1B_1, the angle through which the comet appears to move across the sky in the interval), and its height (SA_1, the distance from the Sun to the comet). If we let $R = SA_1$, then

$$A_1B_1 = R \tan(p)$$

where tan (p) is the tangent of the angle p.

When p is small, as it will be in the present case, then tan $(p) \approx p$ if p is expressed in radians. Then

$$A_1B_1 \approx R \cdot p$$

So, the area of triangle SA_1B_1

$$
\begin{aligned}
&= {}^1/_2 \cdot SA_1 \cdot A_1B_1 \\
&= {}^1/_2 \cdot R \cdot p \cdot R \\
&= {}^1/_2 \cdot p \cdot R^2
\end{aligned}
$$

where p is expressed in radians.

The **radian** is a unit of angle, defined from the condition that the angle around a full circle is 2π radians. Because a full circle contains 360°,

$$1 \text{ radian} = \frac{360}{2\pi} = 57.3°$$

In practice, we will measure angular motion of the comet against the background stars in arc seconds. Because 1° = 60 arc minutes and 1 arc minute = 60 arc seconds,

$$1 \text{ radian} = \frac{(360 \times 60 \times 60)}{2\pi} = 206{,}265 \text{ arc seconds.}$$

With the distance to the comet, R, expressed in AU, the area in units of AU^2 swept out by the comet in one day is:

$$
\begin{aligned}
&= {}^1/_2 \cdot p \text{ (arc seconds)} \cdot (1 \text{ radian} / 206{,}265 \text{ arc seconds)} \cdot R^2 \\
&= 2.42 \cdot 10^{-6} \cdot p \cdot R^2 \\
&= \underline{\text{constant} \cdot p \cdot R^2}
\end{aligned}
$$

Thus, to verify Kepler's second law that the area swept out by the Sun-comet line per unit time is constant, we need to show that the product $p \cdot R^2$ remains constant for Halley's comet along its orbit. To demonstrate this, we will make observations of the comet at several points along this orbit.

We will start by measuring the comet's speed at the point when it is closest to the Sun, at **perihelion**. It is at this position that it will be moving at its fastest.

> 28. Select **Perihelion** from **Go/Observing Projects/Comets**.
> 29. Open the **Info** pane and expand the **Position in Space** layer.

Examine this setup carefully. You are situated at the Sun's north pole at 11:04:40 UT on February 9, 1986, at the moment when Halley's comet passes the precise point of perihelion, shown as a mark on the blue line representing the orbital track of the comet. The field of view is 40 arc minutes wide, and Halley's comet is seen against a background of stars. The Time Interval is set to 1 hour in order to measure the comet's motion at this point in its orbit. The view is locked on to Halley's comet, and thus the background stars will appear to move when time is changed. The comet appears as a small disk in this field of view, which is convenient for the following measurement. (You might want to increase the field of view at some stage to see the foreshortened view of the comet's tail from this viewpoint on the Sun.)

> 30. Note the Sun-comet distance from the **Info** pane. Enter this value under the column R (AU) in the row labeled Perihelion in Data Table 1 below.
> 31. Step the time ahead by one time interval of 1 hour.
> 32. Use the **Hand Tool** to measure the angle between the comet and the perihelion point on the orbital track. Note this angle under the column labeled p_1 in the appropriate row of Data Table 1.
> 33. Step the time back by two 1-hour intervals and again measure the angle between the comet and the perihelion point, noting this value in the data table under the column p_2. This procedure removes any error in setting Halley's comet directly in front of the reference point.
> 34. Calculate the average of p_1 and p_2 and record the result under the column p (' "). Convert this value to arc seconds and enter the result, the angular speed of the comet in arc seconds per hour, in the column p (").
> 35. Calculate the angular speed in arc seconds per day from the above value of arc seconds per hour by multiplying by 24 and enter this value in the data table under the column Angular Speed ("/day).
> 36. Calculate the value of $p \times R^2$ for this position in the comet's orbit and enter this value in the data table.

> **TIP:** If you inadvertently scroll the view while attempting to make this measurement, use the Undo command and attempt the measurement again.

You can now repeat this sequence of measurements for three further positions in the comet's orbit, including the furthest point in its orbit, **aphelion**. The product $p \times R^2$ can be determined in each case, to test Kepler's law.

37. Select **Comet-7AU** from **Go/Observing Projects/Comets.**
38. Adjust the Date to 9/08/1984 A.D. and the time to 14:00:00 UT to place Halley's comet in front of a convenient reference star, TYC742-2450-1.
39. Ensure that the **Info** pane is open and that the **Position in Space** layer is displayed.

Again, examine this setup. We have moved back in time to the date when Halley's comet was at about 6.3 AU from the Sun. In order to see the comet at this time, you are now at the Sun's south pole. The time is 14:00 UT on September 4, 1984, when Halley's comet was passing in front of a star, TYC742-2450-1, from your point of view. The field of view is about 16 arc minutes wide. The time step interval is set to 4 days in order to measure the comet's motion at this point in its orbit. Again, the view is centered on Halley's comet and the background stars will appear to move when time is changed.

40. Record the Sun-comet distance from the **Info** pane in the data table.
41. Step the time ahead by one time interval of 4 days.
42. Use the **Hand Tool** to measure the angle between the comet and the reference star, TYC742-2450-1. Note this angle in the data table.
43. Step the time back by two 4-day intervals to place the comet on the other side of the reference star and again measure the angle between the comet and the star, noting this value in the data table.
44. Calculate the average of these measurements and convert the result to arc seconds.
45. Calculate the angular speed in arc seconds per day from the above value of arc seconds per 4-day period by dividing by 4 and enter this value in the data table.
46. Calculate the value of $p \times R^2$ for this position on the comet's orbit and enter this value in the data table.
47. Select **Comet-14AU** from **Go/Observing Projects/Comets.**

Check this setup. We have moved back in time to April 9, 1981, at 21:00 UT, when Halley's comet was about 14 AU from the Sun. You will note that you are still at the Sun's south pole and the comet appears to be in line with a reference star, TYC749-1294-1, when viewed from this location. The time step interval has been increased to 10 days because the comet will move much slower at this point in its orbit. Again, the view is locked on to Halley's comet and the background stars will appear to move when time is changed.

48. Record the Sun-comet distance from the **Info** pane in the data table.
49. Step the time ahead by one time interval of 10 days.
50. Use the **Hand Tool** to measure the angle between the comet and the reference star, TYC749-1294-1. Note this angle in the data table below.
51. Step the time back by two 10-day intervals to place the comet on the other side of the reference star and again measure the angle between the comet and the star, noting this value in the data table.
52. Calculate the average value of these measurements and convert the result to arc seconds.
53. Calculate the angular speed in arc seconds per day from the above value of arc seconds per 10-day period by dividing by 10 and enter this value in the data table.
54. Calculate the value of $p \times R^2$ for this position on the comet's orbit and enter this value in the data table.

Finally, you can move Halley's comet close to **aphelion**, the farthest point from the Sun in its orbit.

55. Select **Aphelion** from **Go/Observing Projects/Comets.**

At this point in its orbit, Halley's comet is moving much slower. At this distance from the Sun, it is very cold and, even if we could see the comet at this large distance, it would probably show little or no evidence of a tail. Because it is a very small object and we see it only by reflected sunlight, it is invisible to us upon Earth. Fortunately, *Starry Night*™ can easily simulate its appearance for us.

At midnight UT on March 15, 1948, the comet is close to the aphelion point in its orbit, passing in front of a convenient reference star, HIP 40635. The time step interval is now 40 days.

56. Record the Sun-comet distance from the **Info** pane in the data table.
57. Step the time ahead by one time interval of 40 days.
58. Use the **Hand Tool** to measure the angle between the comet and the reference star, HIP 40635. Note this angle in the data table below.
59. Step the time back by two 40-day intervals to place the comet on the other side of the reference star and again measure the angle between the comet and the star, noting this value in the data table.
60. Calculate the average of these measurements and convert the result to arc seconds.
61. Calculate the angular speed in arc seconds per day from the above value of arc seconds per 40-day period by dividing by 40 and enter this value in the data table.
62. Calculate the value of $p \times R^2$ for this position on the comet's orbit and enter this value in the data table.

DATA TABLE 1

View	R (AU)	p_1 (' ")	p_2 (' ")	Average p (' ")	p (")	Interval	Factor	Angular speed ("/day)	$p \times R^2$
Perihelion						1 hour	× 24		
7 AU						4 days	÷ 4		
14 AU						10 days	÷ 10		
Aphelion						40 days	÷ 40		
							Average		

With these data, we can now explore the claim of Kepler's Law as it applies to Halley's comet, that the area swept out by the line between Sun and the comet sweeps out equal areas in equal times. We have calculated the value of $p \times R^2$ for this comet at four points of its orbit. The previous derivation showed that if the product $p \times R^2$ is constant over the range of distances covered by Halley's comet, then we have demonstrated the above law.

You might plot a graph of $p \times R^2$ against R to justify this claim. You will note that this product actually varies systematically with distance, showing that Halley's comet deviates from Kepler's law during its extended journey out into space, with higher-than-average speed at its greatest distance from the Sun. You could explore this deviation from Kepler's law by reading more about comets and their interaction with the interplanetary medium. In practice, the ejection of material from the melting of the comet's nucleus might play a role in altering the comet's speed in its orbit.

Question 11: What is the ratio of largest to smallest distance over which speed measurements were made?

D. Speed of the Comet

It is instructive to calculate the real speeds of this comet as it traverses its orbit, in units of speed commonly used on the Earth, namely km/hour. The product $p \times R$ provides a useful estimate of this speed but needs to be multiplied by a constant to adjust the units to km/hour. The constant is derived below.

$$\text{Orbital speed (km/hour)} = p \ (\text{"/day}) \times 1 \ \text{day/24 hours} \times 1 \ \text{radian/206265"} \times R \ (\text{AU}) \times 1.5 \times 10^8 \ \text{km/AU}$$
$$= 30.30 \times p \ (\text{"/day}) \times R \ (\text{AU})$$

If you now apply this simple formula to your measured results, you will see that this comet is moving very rapidly, particularly when it is close to the Sun. Its impact upon a planet would generate tremendous and devastating energy. The comet Shoemaker-Levy collided with Jupiter in 1992 after being deflected from its orbit and breaking up into more than 20 fragments under tidal stress on a previous close encounter with this giant planet. The impacts of these fragments released enormous energy and disturbed the fluid surface of Jupiter for many months.

> Question 12: What is the orbital speed of Halley's comet at perihelion expressed in km/hour?

> Question 13: What is the orbital speed of Halley's comet at aphelion expressed in km/hour?

E. Conclusions

You have explored the motion and changing appearance of Halley's comet on its last apparition in our sky in 1985 and 1986. You have watched this motion from several locations, some not accessible in real life. You made observations from these unique locations that allowed you to verify that the comet obeys Kepler's second law of planetary motion. From these measurements, you have also been able to calculate values for the tremendous speed that this comet acquires as it sweeps through our solar system and you have demonstrated that it moves much faster when it is near perihelion than at any other time in its orbit.

PARALLAX OF THE SUN AND VENUS

16

The measurement of distance to objects in the universe is crucial to the determination of many other parameters of importance in astrophysics. For example, it is only with knowledge of the distance to a star that we can determine its true output of energy or **luminosity** from the measured intensity of the light that reaches us.

The most direct and straightforward way of determining the distance to a relatively nearby star is the measurement of its **stellar parallax**, i.e., its apparent displacement against the background of more distant objects caused by the movement of the observer around the Earth's orbit. The observed angular displacement is larger if the baseline over which the observations are made is larger. For observers on the Earth, the maximum baseline available is the diameter of the Earth's orbit around the Sun, approximately 2 Astronomical Units (AU).

> The measurement of distance from parallax observations is discussed in Section 19–1 of Freedman and Kaufmann, *Universe*, 7th Ed., and is considered again in the Distance Ladder project.

The parallax of a star, *p,* in arc seconds, is defined as the angle through which the star moves when the observer changes position by 1 AU. The greater the distance to the star, the smaller the parallax. This effect results in a unit of distance, the **parsec**, the distance to an object whose parallax is exactly 1 arc second. The distance, *d,* to any star is then equal to 1/*p*, where *d* is defined in parsecs or pc, and *p* is in arc seconds. In practice, 1 pc is found to be equal to 206,265 AU or 3.26 light-years.

Stars are so distant from the solar system that no stars show greater parallax than 1 arc second, and *Starry Night*™ cannot simulate the effect of stellar parallax. In order to demonstrate the effect of parallax therefore, you can use *Starry Night*™ to examine and measure the effect of parallax on closer objects, particularly the Sun and the planet Venus. For these closer objects, you can use the Earth itself as a baseline, observing their positions against the background stars while moving the observing position in latitude. The extreme brightness of the Sun in particular makes the measurement of its parallax difficult in real life but easy in simulation. A graphical method is used to increase the overall accuracy of the measurements of distances to the Sun and Venus because the precision of angular measurement in *Starry Night*™ is limited in general to 1 arc second.

A. Parallax of the Sun

> The intense brightness of the Sun and the blue scattered daylight makes this observation impossible in real life, but it is simple to simulate with *Starry Night*™.

The parallax of the Sun will be measured by determining the change in the separation between the Sun's apparent position and that of a distant star, while moving the observing position in a north-south direction through a range of latitudes along a constant line of longitude. This longitude has been chosen to place the Sun and star on the local meridian at this time. An observing date and time have been selected when the Sun is passing close to and just north of a reference star, TYC1846-598-1. This geometry provides a baseline equivalent to the length of the Earth's radius.

The diagram below shows a slice through the Earth, coincident with the chosen longitude and with the meridian upon which the Sun and the selected star are to be found. The vertical line labeled A represents the background stars. (E represents the celestial equator.) This is an exaggerated view and, in reality, the background would be very far away from the Earth and the Sun.

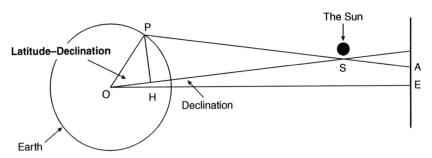

The line OHS from the center of the Earth to the edge of the Sun ends at a fixed point upon the sky, and can be used as a reference line for this simulation. If we now stand at point P and choose a reference star that appears to be near to the Sun's edge, then this star will be at a fixed position with respect to the end of the reference line, OHS. If we now change our latitude in this simulation, we can record the apparent movement of the Sun against the background stars by measuring the spacing between the Sun and the reference star. We can also use the line PH, perpendicular to the reference line OHS, as a reference baseline that changes as latitude changes to produce the apparent movement of the Sun. We can then relate the length of this line to the latitude of the observation point using simple trigonometry.

The latitude, λ, of P is the angle between the Earth's radius OP and the celestial equator OE, shown as a horizontal line in the diagram. The angle between the reference line OHS and the celestial equator is (to within the Sun's angular radius, 0.25°) the declination δ of the Sun in the sky.

Thus, in the triangle OPH in the diagram, the angle POH is simply the latitude of the observer, λ, minus the declination angle of the Sun, δ. That is,

$$\text{Angle } POH = (\lambda - \delta)$$

From trigonometry, the ratio of the side opposite to this angle, PH, to OP, the hypotenuse of the triangle, OP, is the trigonometric Sine function of the Angle. Thus,

$$\text{Sin } (\lambda - \delta) = \frac{PH}{OP},$$

where OP is the radius of the Earth, assumed to be 6378 km, and PH is measured in kilometers. Rearranging gives

$$PH = 6378 \text{ Sin } (\lambda - \delta)$$

The declination of the Sun, δ, can be obtained from the coordinate system or from the Info pane in *Starry Night*™.

Measurements of the angular separation of the Sun from the reference star can now be made from sites at different latitudes and a graph plotted of this separation as a function of Sin $(\lambda - \delta)$. The sine of this angular difference can be obtained from a calculator or the calculator function on your computer. If you now choose an interval over which Sin $(\lambda - \delta)$ changes by 1.0, the above equation shows us that the baseline PH will then change by OP, or the Earth's radius, 6378 km. Thus, from the graph, we can obtain the equivalent change in angle between the Sun and the star, which we can call s, when the baseline changes by this value.

This situation can be represented in a thin triangle, shown below. The angle s, subtended by the baseline $PH = 6378$ km at the Sun in the triangle PSH, is very small and is measured in arc seconds. The distance SH is closely equivalent to the distance OS, the distance from the Earth to the Sun, because OH is very small compared to OS.

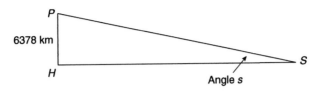

If the distance to the Sun is d km, the following trigonometric relationship is true:

$$\tan s = \frac{PH}{d} = \frac{6378}{d}$$

and, if s is measured in radians and s is very small,

$$\tan s = s = \frac{6378}{d} \text{ (km)}$$

Because 2π radians $= 360°$ or 1 radian $= 57.3°$ by definition, and $1° = 60 \times 60$ arc seconds, s in arc seconds $= s$ (radians) $\times 206,265$.

Rearranging the above equation,

$$d \text{ (km)} = \frac{6378}{s}$$

Because 1 radian $= 206,265$ arc seconds, and 1 AU $= 1.497 \times 10^8$ km, we can rearrange this equation to determine the distance d in AU:

$$d = \frac{6378 \times 206,265}{1.5 \times 10^8} \times \frac{1}{s}$$

$$d = \frac{8.79}{s} \text{ AU, with } s \text{ in arc seconds.}$$

With this background, you can now make a series of measurements to determine the value of s by observing the Sun from sites with different latitudes along a fixed line of longitude, 16° 39' W.

1. Launch *Starry Night*™.
2. Select **File/Preferences**. In the Preferences dialog window, open the **Cursor Tracking (HUD)** page and select only **Name** and **Object type** from the **Show** list.
3. Select the view **Sun-Lat 45S** from **Go/Observing Projects/Parallax**.

You will note that this view, from the center of the South Atlantic Ocean, 45° south of the equator on the chosen line of longitude, shows a magnified view of the Sun at 12:06:00 P.M. on 6/10/2002 A.D. The background stars are visible because the blue daylight sky is suppressed. One of these stars, TYC1846-598-1, has been selected and labeled as a convenient reference star because on this date and at this time from the chosen longitude it is just below the Sun on the local meridian represented by the vertical grey line.

4. Note the approximate declination of the Sun from the coordinate system displayed on the sky (or obtain the exact declination of the Sun from the Info pane under the Position in Sky layer) and record it in Data Table 1 below.
5. Use the **Hand Tool** to measure the angular separation between the center of the Sun and the reference star and record this result in Data Table 1.

You can now repeat this measurement from a series of latitudes, entering the angular separation of star and Sun in the appropriate boxes in Data Table 1.

6. Measure the separation between the center of the Sun and the reference star in each of the views **Sun-Lat 30S, Sun-Lat 0, Sun-Lat 30N, Sun-Lat 45N, Sun-Lat 60N,** and **Sun-Lat 90N** from **Go/Observing Projects/Parallax.** Record these values in Data Table 1.
7. Calculate the difference in angle, $(\lambda - \delta)$, between latitude and the declination of the Sun and use an electronic calculator to calculate Sin $(\lambda - \delta)$. (Be careful about the sign of the difference in angle. For example, when latitude is negative, the difference of latitude and declination, if the latter is also negative, will also be negative.)

DATA TABLE 1

Declination of Sun, δ°				
Location	Latitude, λ°	$(\lambda - \delta)$°	Sin $(\lambda - \delta)$	Angular Separation, Sun to Star
45° S	−45			
30° S	−30			
0°	0			
30° N	+30			
45° N	+45			
60° N	+60			
90° N	+90			

8. On a sheet of graph paper or in a spreadsheet application, plot the angular separation against Sin $(\lambda - \delta)$.
9. Draw the best straight line that fits through the data on this graph. Determine the absolute value of the change in parallax angle when the baseline is changed by 6378 km by measuring the equivalent change in angular position for a change of Sin $(\lambda - \delta) = 1.0$. In practice, this value is simply the absolute value of the slope of the graph. Express the change in angular separation in arc seconds. This will be the value, s, of the parallax.
10. Use the equation $d = \dfrac{8.79}{s}$ to calculate the distance d to the Sun in AU from s, the parallax in arc seconds. Compare this calculated value with the value for the distance to the Sun shown as **"Distance from observer"** under the **Position in Space** layer of the **Info** pane for the Sun.

Question 1:   What is the parallax of the Sun at this time in June 2002?

Question 2:   Using this value, what is the derived distance to the Sun at this time? How does this distance compare to that displayed by *Starry Night*™?

In making these measurements, you have demonstrated the technique of parallax for determining the distance to a relatively close object from its apparent movement against the background caused by the movement of the observer. We can now use the same technique on a somewhat nearer object, the planet, Venus, and determine its distance relatively accurately when it is at inferior conjunction and very close to the Earth.

B. Parallax of Venus

Venus is closer to the Earth at certain times in its orbit than is the Sun, and will thus show a larger parallax shift when the latitude of the observation site is changed. This will occur when Venus is close to inferior conjunction. If a suitable time is chosen when the planet appears to be close to a star in the sky, the parallax will be very apparent. We can select such a time and carry out a sequence of parallax measurements for Venus.

11. Select the view **Venus-Lat 90S** under **Go/Observing Projects/Parallax.**

You will see that, on this chosen date, October 29, 2002 A.D., Venus is moving between the Sun and the Earth and is very close to inferior conjunction. You are looking at the side of Venus that is facing away from the Sun and is therefore in darkness, in its "new" phase. At this close distance to Earth, the planet appears relatively large. This view, from the South Pole in Antarctica, shows the planet less than 2 arc minutes below a reference star, TYC6147-865-1.

You can now follow the same procedure as for the Sun's parallax measurement and examine the planet as the observing position is moved northward along a constant line of longitude. This longitude has been chosen to be 65° 45' W to ensure that the planet and the reference star are on the meridian to simplify the geometry. The time of observation, 1:10 P.M. Standard Time, has been chosen to place the planet directly below the star from the observing sites along this line of longitude.

12. Note the declination of **Venus** from the coordinate system displayed upon the screen.
13. Use the **Hand Tool** to measure the separation of the reference star from the center of Venus, recording it in Data Table 2.
14. To see a graphical illustration of the effect of parallax as you move across the Earth from the South Pole to the North Pole, open the view **Venus-Lat 90N** and note the jump in position of Venus with respect to the reference star. You can alternate these two views several times by using **Edit/Undo "Go" item** and **Edit/Redo "Go" Item** to experience this parallax effect.
15. Open the view **Venus-Lat 60S** from **Go/Observing Projects/Parallax**; measure the angular separation between TYC6147-865-1 and the center of Venus and enter the value in the appropriate box in Data Table 2.
16. Repeat this procedure for the views **Venus-Lat 45S, Venus-Lat 30S, Venus-Lat 0, Venus-Lat 30N, Venus-Lat 45N, Venus-Lat 60N,** and **Venus-Lat 90N,** noting the star-Venus separations in Data Table 2.
17. Calculate the difference in angle, $(\lambda - \delta)$, between latitude and the declination of Venus taking care that, if δ is negative, you account for this in the calculation. (For example, if $\lambda = -50°$ and $\delta = -10°$, then $(\lambda - \delta) = (-50 - (-10)) = (-50 + 10) = -40°$).
18. Use an electronic calculator to calculate Sin $(\lambda - \delta)$, entering the values in the Data Table.
19. Plot a graph of the angular separation against Sin $(\lambda - \delta)$.

Draw the best straight line that fits through the data on this graph and measure the slope of the graph, *s*. The absolute value of the slope will give the value of the difference in angular separation equivalent to a value of Sin $(\lambda - \delta) = 1.0$, and *s* will be the parallax when the baseline is 6378 km.

DATA TABLE 2

Declination of Venus, $\delta°$				
Location	Latitude, $\lambda°$	$(\lambda - \delta)°$	Sin $(\lambda - \delta)$	Angular Separation, Venus to Star
90° S	−90			
60° S	−60			
45° S	−45			
30° S	−30			
0°	0			
30° N	+30			
45° N	+45			
60° N	+60			
90° N	+90			

20. Use the equation $d = \dfrac{8.79}{s}$ to determine the distance from the Earth to Venus in AU and compare this to the value displayed in the Position in Space layer in the Info pane.

Question 3: What is the parallax of Venus for a baseline of 6378 km on this day in October 2002 A.D.?

Question 4: Using this value of the parallax, what is the derived distance from the Earth to Venus at this time? How does this compare to the distance to Venus, displayed when the Hand Tool is over the planet?

C. Conclusions

In this project, the parallax of the Sun and Venus against the background stars has been measured as a function of the observer's latitude. These measurements have been used to deduce the distance of these objects from Earth, using a similar method to that used for the measurement of distances to nearby stars when the baseline is the Earth's orbit.

PROPER MOTION OF STARS 17

Stars are not fixed in the sky. They move relative to one another, but the motion is very slow compared to that of the Moon and planets. For example, thousands of years would pass before we would notice significant changes in the shapes of the familiar constellations. Thus, over a person's lifetime, to the unaided eye the stars appear to remain fixed in position. It is only by using telescopes and precise measurement techniques, or by comparing measurements or photographs taken many years apart, that we can detect the relative movement of stars.

The motion of stars across the sky relative to each other, or, more correctly, relative to a fixed coordinate system in the sky, is referred to as **proper motion**. "Distances" across the sky are always measured as angles because the sky has the appearance of a perfect sphere (the celestial sphere) at an infinite distance from us. A complete circle around the sky is 360°, so a small distance across the sky is some part of 360°, expressed as an angle in degrees (°), minutes of arc ('), and seconds of arc ("). In view of the scale of stellar movement, the usual units for proper motion are seconds of arc per year ("/yr).

Sir Edmund Halley discovered proper motion in 1718. Halley, for whom Halley's Comet is named, compared the positions of stars in his time to their positions 1500 to 2000 years earlier as measured by the ancient Greek astronomers Ptolemy and Hipparchus, and realized that the stars Sirius, Aldebaran, and Arcturus had moved slightly relative to the other stars.

The discovery and measurement of proper motion of other stars became far easier after the invention of photography and its application to astronomy in the late nineteenth century. With this technique, it became possible to compare simultaneously the positions of many stars on two photographs taken several years apart.

In 1916, the American astronomer E. E. Barnard discovered the star with the highest known proper motion, a faint star in the constellation Ophiuchus. This star is now known as

> The proper motion of Barnard's star is shown in a composite photograph in Fig.19–3 of Freedman and Kaufmann, *Universe*, 7th Ed.

Barnard's star and at the present time is situated 5.94 LY (light-years) from the Earth. It is a small star, having a mass and a diameter of about one-sixth of that of the Sun and an apparent magnitude in our sky of +9.5, too faint to be seen with the unaided eye. It is traveling toward the Sun and will pass within 3.8 LY from us somewhere around 10,000 A.D. For comparison, the nearest known stars to the Sun at the present time are those in the Alpha Centauri triple system, at a distance of about 4.3 LY.

In this project, we will investigate the proper motions of Barnard's star and other stars.

A. Observing the Proper Motion of Barnard's Star

We first set up a convenient date, time, and location to observe the proper motion of this rapidly moving star. You can watch Barnard's star as it traverses a significant region of our sky over centuries of time. We will then choose a specific time to observe its close approach to another star in order to measure this proper motion accurately.

1. Launch *Starry Night*™ and select the view **pmotion-A** from **Go/Observing Projects/Proper Motion of Stars.**

HIP87946 is the name of a faint star in the constellation Ophiuchus. The name identifies the star as being the 87,946th entry in the Hipparcos Catalog, a listing of 120,000 stars whose positions and distances were measured very accurately by the European Space Agency's Hipparcos satellite.

The Tycho Catalog, a companion listing produced from data obtained with the same satellite, lists 400,000 stars whose positions and distances were measured somewhat less accurately. These stars have names beginning with the prefix "TYC."

This initial field of view is about 3° across and shows a rich field of stars as seen from the North Pole of the Earth at midday on October 14, 1500 A.D. This site is chosen to avoid having the horizon interfere with observations. The field is centered on the position of a faint star, HIP87946, which is too faint to be displayed by this version of *Starry Night*™, but because its position is marked, it can still be used as a marker against which you can watch the motion of Barnard's star.

2. Run **Time Forward** and watch the region of the sky near the reference star as time advances in 2-year steps.

With the view locked on to the position of the star HIP87946, this star and most of its neighbors appear to remain fixed as time advances. However, you will note that one star moves at a very noticeable rate upward in the view during this time, passing close to the reference star in about 1844 A.D. This is Barnard's star, showing a very high proper motion.

3. Select the view **pmotion-A** from **Go/Observing Projects/Proper Motion of Stars** to watch the motion of Barnard's star again. Allow time to continue past the point at which Barnard's star exits the top of the view and observe the other stars carefully.

You will notice three things.

First, Barnard's star moves very rapidly across the sky, demonstrating a very high proper motion.

Second, in this small region of the sky spanning a few degrees in width, several other faint stars show noticeable movement with respect to their neighbors over a long time period. You should search for these slowly drifting stars.

Third, the whole sky is slowly rotating as the Earth's spin axis precesses. Because the coordinate system is linked to the Earth's spin axis, our view will appear to rotate very slowly. This motion is periodic with a repetition time of about 26,000 years and is discussed in the Precession project.

You can now use the fact that Barnard's star passes very close to the position of the reference star HIP87946 around 1844 to measure its proper motion at around that time.

B. Measuring the Proper Motion of Barnard's Star

Proper motion of a star is defined as a **magnitude** of motion in arc seconds per year and a **direction** with respect to the equatorial coordinate system, usually quoted as an angle in degrees from north, measured in an easterly direction.

The magnitude of Barnard's star proper motion can be measured by setting a time and date when this star is close to a reference star. Stepping back and forward in time will allow you to measure the motion of Barnard's star with respect to the reference star over a specific time interval. To improve the accuracy of this measurement, the chosen time interval is relatively long. The sky will rotate over a long period because of precession but if the view is centered and locked upon the reference star this rotation will not affect the measurement of spacing between the two stars.

4. Select the view **pmotion-B** from **Go/Observing Projects/Proper Motion of Stars**.

This view shows a 1° field surrounding the reference star at midnight on August 8, 1844 A.D. from the North Pole of Earth. At this time, Barnard's star is at its closest to, and within 2 arc seconds of, the reference star HIP87946. You can now adjust the time and watch the close star move with respect to the more distant star and use the Hand Tool to measure the extent of this movement over a specific time interval, ahead of and before the date of closest approach.

5. Step the **Time Forward** by 2 time intervals of 100 years.
6. Use the **Hand Tool** to measure the angular spacing between the position of the reference star and Barnard's star and note the value in Data Table 1.
7. Step **Time Back** by four 100-year time intervals to carry Barnard's star to the other side of the reference star.
8. Use the **Hand Tool** to measure the angular spacing between Barnard's star and the position of the reference star and note this value in Data Table 1.
9. Translate your measurements to arc seconds and enter them in the appropriate box of the data table.
10. Add the two measurements together and divide the result by 400 to determine the magnitude of the proper motion of Barnard's star in arc seconds per year.

DATA TABLE 1

Time step from reference time	Spacing between stars (', ")	Spacing (")
+200 years		
−200 years		
	Motion in 400 years	
	Proper motion ("/year)	

Question 1: What was the magnitude of the proper motion of Barnard's star in arc seconds per year in 1844 A.D.?

The direction of this proper motion can be measured with respect to the Right Ascension-Declination coordinate system but in this case, the measurement must be made over a much shorter time interval to avoid the rotation of the field of view that occurs over longer times because of precession. Thus, we will zoom in to a much smaller field of view and change the time step to 10 years to measure the angle of this motion with respect to the north direction. We can then use simple trigonometry to determine the angle that the star's proper motion makes with the N-S line, usually quoted in the direction of east from the North direction.

11. Select the view **pmotion-B** from **Go/Observing Projects/Proper Motion of Stars**.
12. Adjust the field of view to 6 arc minutes and change the **Time Step** to 10 years.
13. Select **View Options/Guides/Celestial Grid** to overlay a north-south grid upon this region of the sky.
14. Step **Time Forward** by one 10-year time interval.

Barnard's star has now moved northward and a little to the west, as you will see by the fact that the spacing between it and a N-S coordinate system line is now larger than that between the reference star and the same N-S line. We can use this change to measure the angle of the proper motion. (You might have noticed that the right ascension shown on the N-S line next to HIP87946 after you advanced time is not the same as the one before you advanced time because the entire coordinate system has shifted to the west, i.e., to the right on the screen, due to precession of the Earth's coordinate axis. However, we are using these lines only to establish the direction of north, not the actual coordinates of the stars, and the direction these lines point [N–S] has changed very little over 10 years.)

The following diagram shows the geometry after this time advance. O shows the position of the reference star, **B** is the position of Barnard's star, and the vertical line AO is the N–S direction of the coordinate system. The position angle for the proper motion is always stated as the angle from the north direction toward the East, as shown in the diagram. This angle is simply 360° minus the small angle AOB. We can determine AOB using measurements of the angular separations AB and OB, which for small angular separations can be considered as distances.

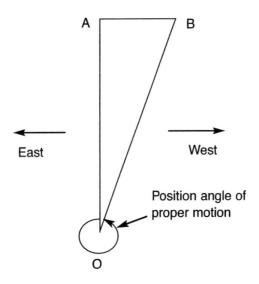

In triangle AOB, the ratio of AB to OB is the trigonometric Sine of the angle AOB, or

$$Sin\ (AOB) = AB/OB$$

To determine AB, measure the spacing between Barnard's star and a convenient N–S line of the coordinate system and the equivalent spacing between the position of the reference star and the same N–S line. The difference between these measurements is the angular spacing AB.

To determine OB, measure the spacing between the reference star and Barnard's star.

15. Use the **Hand Tool** to measure the spacing between Barnard's star and a convenient N-S coordinate line and note this value. Repeat this measurement between the position of the reference star and the same N-S line and note this value. Calculate the difference of these two values, AB arc seconds.

16. Use the **Hand Tool** to measure the spacing OB between Barnard's star and the reference star, and calculate this angle in arc seconds.

17. Calculate the ratio of these two angles, AB/OB, and use a calculator or the calculator function of your computer to obtain the angle AOB by taking the inverse sine of this ratio.

18. Calculate (360° – Angle AOB), the position angle of the proper motion of Barnard's star.

Question 2: What was the position angle of the proper motion of Barnard's star, measured east from the North direction, in 1844 A.D.?

C. The Proper Motion of Other Stars

The proper motion of other stars is so much slower than that of Barnard's star that we will need a much longer time step in order to see this motion clearly on the screen. We can watch sky motions over the full time span of about 14,712 years available in *Starry Night™*. However, because we have positioned ourselves above one end of the Earth's rotation axis (the North Pole) in the previous sequences, precession makes our view of the sky change with time. This motion of the sky partially masks the effects of proper motion. We can remove this effect by positioning ourselves at the north pole of the Sun, something that is impossible to do in real life but simple to achieve in simulation.

> 19. Select the view **pmotion-C** from **Go/Observing Projects/Proper Motion of Stars**.
> (*N.B.:* The Location in the Location Box does not always appear to change in this step but the location is now above the North Pole of the Sun.)

The view is still centered upon the same reference star in this 5°-wide field, and the time is set to the earliest date available in *Starry Night™*.

> 20. Set **Time** running in 100-year steps and watch the stars as they demonstrate their individual proper motions across the sky. If this motion is too slow or too fast, adjust the Time Interval to a more suitable value.

Notice that stars appear to move in every possible direction on the screen with many diverse speeds. Several stars move very quickly while others appear not to move at all. (HIP87946 does not move because we have locked the screen on to it. However, it is also very far away, and as a consequence, its proper motion is very small.)

If you watch very carefully, you may see Barnard's star flying past HIP87946 between about 1300 A.D. and 2400 A.D., moving almost vertically through the center of the view. (If not, try 50-year time steps, starting at 1300 A.D.) The program stops at 10,000 A.D. To repeat this display, simply reverse the direction of time change.

There is one other fascinating situation that shows up in this long time-span. The measurements of the individual motions of a binary pair of stars orbiting each other have been misinterpreted as proper motions and the simulation assumes that these components are moving with their individual motions across the sky.

> 21. Select the file **pmotion-C** again from **Go/Observing Projects/Proper Motion of Stars**.
> Identify the relatively bright star 70 Ophiuchi to the east and north of the reference star. Observe this star as time progresses forward. You will see that another star converges upon it until, at around the present time, they appear as one star. If you run time back again to separate these two stars, you will note that they both have the same name, 70 Ophiuchi. They are members of a binary star but simulation using their mutual motions as proper motions has shown them to be moving apart.

D. Relationship Between Proper Motion and Distance

You can explore the possible relationship between observed proper motion of a star and its distance from the Earth with the small sample of stars in this limited field of view.

> 22. Select the view **pmotion-C** again from **Go/Observing Projects/Proper Motion of Stars** to return to the view of the stars in the constellation Ophiuchus.

23. Position the cursor over any star and use **Show Info** option in the contextual menu to open the **Info** pane. Expand the **Position in Space** layer in this pane.
24. Run **Time Forward** and **Backward** and watch for several stars that show rapid motion over this time-span.
25. Stop time flow and click on several of these high-speed stars in turn and note their respective distances from the Earth as given in the Info pane.
26. Run **Time Backward** and **Forward** several times to look for stars that appear to move relatively slowly over the time interval.
27. Stop time flow and click on several of the stars that move much slower and again note their distances from the Earth.
28. Finally, run **Time Backward** and **Forward** again to repeat the sequence and watch for stars that show no motion at all over this timespan.
29. Stop time flow and click on these relatively stationary stars and note their distances from the Earth.
30. On the basis of this small survey, answer the following general questions about the relationship between distance from the Earth and proper motion.

Note that the stars with the most rapid proper motion, apart from Barnard's star, are the two components of 70 Ophiuchi, examined above. Unfortunately, no reliable distance has been assigned to these stars and they cannot be used in your survey.

Question 3: From your survey, what approximate relationship do you see between the proper motion of stars in the solar neighborhood and their distances from the Earth?

Question 4: Why do you think that this relationship exists?

31. Exit *Starry Night*™ without saving changes to the file pmotion-C.

E. Conclusions

In this project, you have observed and measured the proper motion of our Sun's fastest-moving neighbor, Barnard's star, as it moves through our sky. You have also been shown a glimpse of the complex motions of a small sample of stars near to our Sun as they move in space over many centuries.

THE HERTZSPRUNG–RUSSELL DIAGRAM

18

THE CLASSIFICATION OF STARS

The study of stars occupies a very important place in the overall study of our universe. This is at least partly because the source of most of the energy that we need for our survival on this Earth comes directly or indirectly from one such star, our Sun!

We learn about stars by analyzing the light and other forms of radiation, such as ultraviolet light, X rays, and infrared radiation that comes to us across vast distances. This radiation originates in the outer layers of each star, at its so-called surface, or in its atmosphere. The interior of the star, where most of its mass resides and where all its energy is generated, is hidden from our view. When we analyze the light from these outer layers, we find that most stars are very similar to one another in terms of their composition. Almost all stars, at least in their outer layers, are composed primarily of the very simple element hydrogen, with a relatively small fraction of helium and a very much smaller fraction of heavier elements. (Astronomers refer to all elements heavier than hydrogen and helium as "metals.") In terms of their observed composition, it is only this very small metal abundance that varies significantly from one star to another.

Given this great similarity in their observed composition, it is perhaps surprising that stars can take on such a wide variety of forms, from large to small, bright to faint, and hot to cool. It would seem from this that the composition of the surface layers is not the major factor that determines the overall characteristics of a star. The wide array of forms must arise from differences in the interior structure of stars. Also, the simple fact that stars radiate copious amounts of energy means that they must change with time; i.e., stars may "evolve" from one form into another.

To understand these forms and how they might be related to each other, astronomers more than a century ago began to classify stars into categories. This classification scheme, and the desire to explain why stars fall into these categories, then prompted scholars to construct theoretical models of stars using the laws of physics and known properties of matter, such as nuclear and atomic interactions, the emission and absorption of radiation, and the compression of matter by gravity. Present-day theoretical models have developed to the point where they can predict the structure and evolution of stars from birth to death with reasonable accuracy. The final test of the theory, however, is that the computed stellar models must match the observed properties of actual stars. Thus, it is useful to consider these properties and their classification in a way that is used by professional astronomers. This project outlines this process of classification, and relates it to various physical properties of stars.

Several easily measured **stellar parameters** are used in the classification schemes of stars that were devised early in the history of astronomy. **Stellar brightness** at certain well-defined colors has been measured using the techniques of photometry, initially using visual observations, then photographic imagery and finally photoelectric measurements. These brightnesses are related to the star's energy output and are expressed by astronomers in the logarithmic scale of **magnitudes**. They are referred to as **apparent magnitudes** when they represent measurements made from Earth and are usually assigned the letter "m," with a suffix denoting the color at which the measurement was taken.

Section 19-3, Figure 19-6, and Box 19-2, Tools of the Astronomer's Trade, in Freedman and Kaufmann,*Universe*, 7th Ed., discuss and illustrate the modern magnitude scale.

Magnitude values increase as brightness decreases in a way that has more to do with history than utility. Stars were assigned a "magnitude" by early Greek astronomers, the brightest being of "first magnitude," the faintest of "sixth magnitude." Astronomers still use a modern version of this inverted logarithmic scale even though it is inconvenient compared to discussing actual brightnesses on a linear scale. In this scale, a magnitude change of 1.0, for example from magnitude 1.0 to magnitude 2.0, represents a brightness decrease of a factor of 2.512. Thus, a change from 1st magnitude to 6th magnitude is 5 magnitudes difference, which translates to $2.512 \times 2.512 \times 2.512 \times 2.512 \times 2.512 = 100$ to match approximately the ratio of the brightness of 1st to 6th magnitude stars as defined by the early Greek observers.

Sections 19-2 and 19-4 and Figure 19-7 in Freedman and Kaufmann, *Universe*, 7th Ed., discuss and illustrate photometry and its relationship to a star's surface temperature.

Photometry at different wavelengths can also be used to define a stellar **surface temperature** if the star is assumed to emit approximately as a blackbody, an assumption that is reasonable for most stars.

Spectroscopy, the detailed examination of the spectrum of stars upon which a complex pattern of dark absorption (and sometimes bright emission) lines appear, also provided a convenient classification method in early studies. Stars with different sets of absorption lines in their spectra were assigned different letters. When it was discovered that the different types of absorption-line spectra were related to stellar surface temperature, these letters were placed in a sequence, O, B, A, F, G, K and M relating to decreasing temperature.

Sections 19-5 Figure 19-11 in Freedman and Kaufmann, *Universe*, 7th Ed., discuss and illustrate stellar spectral classification.

Measurement of **distances** to stars, for a long time a difficult task, began to provide the capability to estimate the absolute brightness of stars by adjusting the apparent brightness for distance via the **inverse square law**. This led to **absolute magnitudes**, designated by the letter "M," again with a suffix appropriate to the color at which this brightness measure is defined. Absolute magnitude is the magnitude that a star would have if it were at a **standard distance of 10 parsecs** from Earth and therefore represents a measure of star output independent of distance from the observer.

Sections 19-2 and 19-3 in Freedman and Kaufmann, *Universe*, 7th Ed., discuss absolute magnitude, luminosity, and the inverse square law. The relationship between absolute magnitude, apparent magnitude, and distance are discussed in Boxes 19-2 and 19-3.

These brightness values at certain colors could then be adjusted by using the overall shape of the blackbody spectrum of the star derived from the temperature to provide an estimate of the total energy output of the star, its **luminosity**. This measure of total energy output of the star, usually represented in terms of the luminosity of the Sun, is an important parameter for comparison with predictions of theoretical stellar models.

Chapters 20, 21, 22, and 23 of Freedman and Kaufmann, *Universe*, 7th Ed., discuss stellar evolution.

Finally, there is a relationship between **luminosity, L, temperature, T,** and the star's size represented by its **radius, R**. A blackbody at a certain temperature is known to emit a certain amount of energy per unit area per second, E, that is related to its temperature by $E = \sigma T^4$, where σ is a constant. Thus, a star of radius R with a surface area of $A = 4\pi R^2$ will emit a total amount of energy per second represented by its luminosity $L = E \times A$, or

Box 19-4, Tools of the Astronomer's Trade, of Freedman and Kaufmann, *Universe*, 7th Ed., discusses the equations $E = \sigma T^4$ and $L = 4\pi R^2 \sigma T^4$.

$$L = 4\pi R^2 \sigma T^4$$

Early in the last century, two astronomers, Ejnar Hertzsprung in Denmark and Henry Norris Russell in the United States, began to experiment with ways to represent stars collectively in terms of these observed and derived parameters. They independently hit upon a graphical method that has become the keystone, not only of classification of stars, but also of the study of the evolution of stars with time.

Section 19-7 and Figure 19-4 in Freedman and Kaufmann, *Universe*, 7th Ed., discuss and illustrate the H-R diagram. The colors in Figure 19-4 represent the approximate colors of stars in the various regions of the diagram.

The **Hertzsprung–Russell diagram,** or H–R diagram as it is often called, is a scatter diagram of the luminosity of stars plotted as a function of their surface temperatures. In view of the wide range of these parameters for stars, these scales are usually plotted with non-linear scales. Luminosity is plotted logarithmically along the vertical axis, while the horizontal axis shows either temperature plotted logarithmically, or the spectral letter classifications O, B, A, F, G, K, and M plotted equally spaced. (The latter actually corresponds quite closely to the logarithmic temperature scale.) Other versions of the H–R diagram may denote absolute magnitude of the stars against their spectral classification. All H–R diagrams plot temperature in an inverse direction, with high temperatures on the left and low temperatures on the right.

There are several characteristics of note in this scatterplot of stellar parameters. Stars tend to group together in certain areas of the plot, whereas other areas are almost devoid of stars. The majority of stars are concentrated in a diagonal sweep across the diagram from high temperature-high luminosity to low temperature-low luminosity. This region is known as the **main sequence.** This is the region where theoretical models predict that stars will congregate when they are "burning" hydrogen in their interiors in nuclear reactions. The lower edge of this region is known as the **zero-age main sequence** (ZAMS) to designate the line where stars of different mass first begin to burn hydrogen in their cores. This is the locus of points for stars that have reached the time in their lifetimes when pressure generated by the heat of nuclear burning is sufficient to balance the gravitational force that has been condensing the pre–main-sequence star.

Few stars are found below the ZAMS. However, those that are there form a class of very hot stars that, despite being very hot, are so small that their luminosity is very small. These **white dwarf stars** represent old and very evolved stars that have shed their outer layers to reveal a very small but extremely hot inner core. These stars are no longer generating energy but are merely emitting light as they cool.

Stars with high luminosities but relatively low temperatures occupy a wide region above the main sequence. Again, theoretical prediction indicates that the majority of these are stars that have consumed all the hydrogen in their cores and have expanded and cooled as a result of internal readjustment, but are still burning helium and other elements. These are the **red giants.**

There are stars with enormous outputs of energy represented by very high luminosities that cannot be very old because they are burning their fuel at a prodigious rate. These are the **supergiants,** and they can be hot or cool, hence blue or red in color.

The final stellar parameter that is important in this work is the **mass** of the star. This is another parameter that is difficult to measure but which is vital to this modeling. **Stellar mass** represents the amount of fuel that is available for nuclear burning. However, it is also found that the larger the initial mass of the star, the hotter the star's core, thereby accelerating the burning process and significantly affecting the resulting luminosity. The energy output of a higher-mass star is so great that, despite the larger amount of fuel, it uses up its fuel in a shorter time than does a lower-mass star. Because of this effect, we find a relationship between the **mass** of a main-sequence star and its **luminosity** and its **age.** The larger the mass, the greater the output of energy and the shorter the lifetime of the star. This in turn leads to the conclusion that the more massive stars at the high-temperature high-luminosity end of the main sequence are young and will evolve rapidly away from this position. The mass-luminosity relation, shown as a graph of luminosity *vs.* mass later in this chapter, will be used for determining masses of selected stars from an observed list.

Section 19-19 of Freedman and Kaufmann, *Universe*, 7th Ed., discusses the mass-luminosity relation for main-sequence stars.

Two further concepts concerning the H–R diagram are interesting. The first concerns regions on this diagram where no stars appear. There are two reasons why this can be so. First, conditions may never be suitable to produce a star with these luminosities and temperatures. Second, a star can possess these values of luminosity and temperature, but its motion through this region in its evolution is so rapid that the probability of finding a star in this position is small. Theoretical modeling of stellar evolution can predict a star's path across the H–R diagram (and thus delineate the regions of the H–R diagram that cannot be occupied by stars) and the speed at which it follows this path. The observed number of stars in the different regions of the H–R diagram therefore provides another important test of the theory.

Figures 20–17 and 20–18 in Freedman and Kaufmann, *Universe*, 7th Ed., show the H–R diagrams for young clusters in which the lower-mass stars are still evolving toward the main sequence.

Figures 21–11 and 21–12 in Freedman and Kaufmann, *Universe*, 7th Ed., show the H–R diagrams for a globular cluster and a compilation of open clusters, respectively. Both figures illustrate the turnoff points for the clusters.

The second concept concerns the classification of stars in a **cluster** of stars. A cluster forms from a single interstellar cloud, or even from just part of a single interstellar cloud if it is a large cloud, so we can be reasonably sure of three conditions. First, an interstellar cloud has a relatively uniform composition, so the stars in the cluster form with the **same chemical makeup**. Second, we can assume that all of these stars formed at about the **same time**. Third, the **differences** in observed star properties are then caused by differences in the **initial mass** that came together to form each star. This makes the study of clusters important for the verification of evolutionary models. The more rapid evolution of the massive and highly luminous stars at the top end of the main sequence means that such stars will be absent for an old cluster. Indeed, the position of the upper end of the remaining main sequence on the H–R diagram of a cluster, the so-called **turnoff point**, is a good indication of the age of the cluster.

In this project, you will examine a series of chosen stars, but instead of having to go through the tedious process of measuring the various parameters of these stars and classifying them, you can find them in the sky with *Starry Night™* and use the Info pane to obtain the data for each star in turn to build up your own composite H–R diagram.

Thus, you will be able to classify these stars and maybe reach some conclusions about their past and future by comparing their positions on your H–R diagram with the results of stellar evolution modeling.

A. Classification of Stars on an H–R Diagram

The list in Data Table 1 contains many bright stars in our sky, with a wide range of properties. The procedure will be to find each star in turn using *Starry Night™*, retrieve the necessary information from the Info pane for each star and tabulate the information in Data Table 1. You can then plot these parameters on the H–R diagram graphical template and classify the stars into the different groups discussed earlier. The parameters that are most easily plotted on this graph are absolute visual magnitude, M_v, on the vertical axis because this brightness representation is linear but on an inverted scale, against temperature on the horizontal scale. The temperature scale is reversed and nonlinear but the intervals of temperature are marked such that points can be plotted easily.

1. Launch *Starry Night™*.
2. Select the view **HR Diagram-A** under **Go/Observing Projects/HR Diagram**.
3. For each star listed in Data Table 1, open the **Find** pane, enter the name exactly as it is written in the data table, and press Enter. You will see the sky slew rapidly to center the chosen star in the main view window.
4. Select **Show Info** from the contextual menu for this star to display a table of information. You may need to magnify the image by reducing the Field of View in order to identify the star correctly, particularly when that star is in a cluster (e.g., Merope, in the Pleiades) or is a member of a binary pair of stars (e.g., Alpha2 Centauri, with Rigil Kentaurus within 13 arc seconds of it).
5. Expand the **Other Data** layer, and enter the values of **absolute magnitude, temperature, luminosity,** and **radius** for this star into **Data Table 1**. These quoted values for luminosity and radius have been derived from the absolute magnitude and temperature. You will not need the luminosity or the radius for plotting these stars on an H–R diagram. However, you will need the luminosity for determining the mass of a selected set of main-sequence stars in a later section of this project. You will also see that the H–R graph template shows a set of diagonal coordinate axes for radius to show the mutual relationship between luminosity, temperature and radius, as described above.
6. If you wish, you can choose other stars at random to fill out your H–R diagram, or add your favorite star to this graph.

TIP: In step 3, after slewing starts, press the space bar on your keyboard to go there instantly.

Figure 1 is an H–R Diagram template that you can use to plot the absolute magnitude of each star against its temperature. Also shown on Figure 1 are the approximate positions of the main groups of stars, and a solid line representing the approximate position of the main sequence.

7. In **Figure 1**, plot the **absolute visual magnitude, M_v** (Column 2), of each star in Data Table 1 vertically against the **temperature** (Column 3) horizontally. Note that the M_v scale is inverted, with M_v increasing downward and that the temperature scale is reversed and non-linear.
8. In the column labeled "Type of Star" in **Data Table 1,** note the type of star as derived from its position in Figure 1 with respect to the areas labeled main sequence, giant, supergiant, etc.

White dwarf stars are either too faint or too close to bright companion stars for *Starry Night*™ to display them. To fill in this gap, Data Table 2 contains the parameters of a few representative white dwarf stars for inclusion in your H–R diagram.

9. In Figure 1, plot M_v against temperature for the three white dwarf stars listed in **Data Table 2.**
10. Estimate the luminosity of each white dwarf star in terms of the Sun's luminosity from the scale on the right-hand side of the H–R diagram and enter your estimates in Data Table 2.
11. Estimate the radius of each white dwarf star in terms of the Sun's radius from the diagonal radius lines labeled in solar radii across the H–R diagram and enter your estimates in Data Table 2. Again, note that the scale of this oblique Radius scale is logarithmic and your estimates will be somewhat approximate as a consequence.

Question 1: Are there regions on your H–R diagram that appear to be devoid of stars? If so, where are they (e.g., high-temperature/high luminosity, low temperature/high luminosity, etc.)?

Question 2: Are these regions also devoid of stars in more extensive H–R diagrams displayed in your textbook or other astronomy texts (e.g., The Hipparchos data displayed in Figure 21-8[b] of Freedman and Kaufmann, *Universe*, 7th Ed.)?

Question 3: Based on the three white dwarf stars in Data Table 2, what is the luminosity of a typical white dwarf star in terms of the Sun's luminosity?

Question 4: What are the radii of the three white dwarf stars in Data Table 2?

Question 5: Into which classification group—white dwarf, main sequence, giant, or supergiant—would you place the star Deneb?

DATA TABLE 1

1 Star Name	2 Absolute Visual Magnitude, M_v	3 Temperature (K)	4 Luminosity (solar L)	5 Radius (solar radii)	6 Type of Star	7 Mass (solar masses)
Vega						
Aldebaran						
Sirius						
Capella						
Procyon						
Rigel						
Betelgeuse						
Bellatrix						
Mintaka						
Merope						
Miaplacidus						
Zeta Ursae Majoris						
Fomalhaut						
Altair						
Beta Pictoris						
Alpha 2 Centauri						
Mu Cassiopeiae						
Castor						
Sadr						
Altais						
Sarin						
Theta Lyrae						
Spica						
Deneb						
Canopus						
Polaris						
Epsilon Eridani						
Arcturus						
Pollux						
Antares						
Hadar						
Alpha Lupi						

DATA TABLE 2

Star Name	Temperature (K)	Luminosity (solar L)	Absolute Magnitude	Type of Star	Radius of Star (in solar *R*)
Sirius B	26,000		+11.5	White dwarf	
40 Eridani B	15,000		+11.0	White dwarf	
Procyon B	8,200		+13.0	White dwarf	

FIGURE 1: H–R Diagram Template

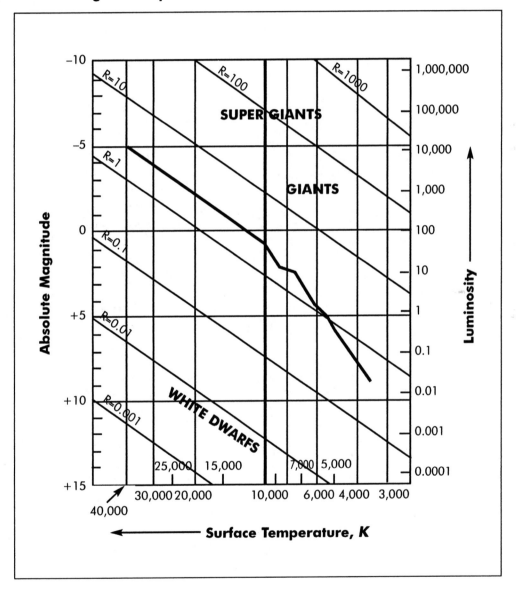

B. Masses of Main-Sequence Stars

In the previous section, you were able to classify a list of bright stars on the H–R diagram and identify several main-sequence stars among them. In this section, you now use the well-known relationship between the luminosity and mass of this class of stars to determine their masses. This important physical parameter of stars is difficult to measure directly. In fact, the direct measurement of a star's mass depends upon its gravitational influence upon a companion star in a binary pair and can be made only for a few nearby stars where the properties of the binary can be measured. Thus, the mass-luminosity relation for main-sequence stars, verified for a few stars by direct measurement, is very important in providing estimates of mass for a much wider range of stars.

The theoretical basis for this relationship comes from the idea that the more massive the star, the more intensely the nuclear furnace will burn the hydrogen in its interior, producing a higher energy output from the star; i.e., a higher luminosity. The graph shown in Figure 2, below, is a summary of stars with a range of masses between 0.1 and 20 solar masses. Densities and temperatures of hydrogen in the interiors of stars with masses lower than about 0.1 solar masses will never be sufficient to trigger nuclear reactions, and such stars will merely emit energy that has been generated by the contraction of the gas by gravity. Because of the absence of nuclear reactions, these objects are not true stars, and are usually called brown dwarfs. Above a higher mass limit of about 80 to 100 solar masses, stars become extremely unstable, the nuclear furnace burning so vigorously that stars rapidly reach an explosive and destructive stage.

FIGURE 2: Mass–Luminosity Relation

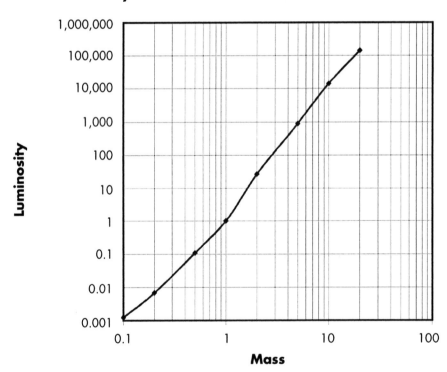

As you will have determined from Figure 1, the following list of stars are considered to be main-sequence stars and their masses can be determined from the graph of luminosity against mass, the so-called Mass-Luminosity Relation. Notwithstanding the fact that both axes in this graph are logarithmic in scale, you should be able to obtain a reasonable estimate of the mass of each of these stars.

MAIN-SEQUENCE STARS

Low Temperature	Medium Temperature	High Temperature
Alpha Centauri Mu Cassiopeiae	Procyon Fomalhaut Altair Beta Pictoris	Bellatrix Spica

12. For each of the stars in the table labeled **Main-Sequence Stars**, above, use the value of **luminosity** obtained from *Starry Night*™ to estimate its mass, by entering the listed value of **luminosity** on the vertical axis of **Figure 2** and reading off the associated value of **stellar mass**, in units of the solar mass, taking careful note of the fact that both axes of this graph have logarithmic scales. Enter this mass in the appropriate box of **Data Table 1.**

You will note that the range of masses of these stars is very significantly smaller than the range of luminosities. Indeed, the above relationship can be represented approximately by a power law, in which luminosity is proportional to mass to a high power,

$$L = kM^{3.6}$$

where k is a constant. This equation shows that stars with large masses will have very much larger luminosities than their lower-mass counterparts.

Section 19–9 in Freedman and Kaufmann, *Universe*, 7th Ed., discusses the determination of stellar masses and the mass-luminosity law for main-sequence stars.

C. Average Density of Stars

Now that you have determined the masses of several main-sequence stars and obtained values for their radii, it might be interesting to determine the average density of the gases in these stars and compare it to a typical density on Earth, such as that of water, whose density is 1000 kg/m³.

Average density is simply the ratio of the mass, M_S, of the star over its volume, V_S, which is given by $4\pi R_S^3/3$. Because the mass and radius of these stars are given in terms of solar mass M_0 and solar radius R_0, respectively, the mass and radius of these stars in physical units of kg and m are $M_S = M \times M_0$ and $R_S = R \times R_0$, respectively. Thus, we can write

$$\text{Average density } \rho = \frac{M_S}{V_S} = \frac{M \cdot M_0}{\dfrac{4\pi(R \cdot R_0)^3}{3}} = \left[\frac{M_0}{\dfrac{4\pi R_0^3}{3}}\right] \cdot \left[\frac{M}{R^3}\right]$$

The first term on the right-hand side of this equation is simply the **average density** of the Sun, which can be calculated to be $\Delta_0 = 1410$ kg/m³, using the values

$$R_0 = 1 \text{ solar radius} = 6.96 \times 10^8 \text{ m}$$
$$M_0 = 1 \text{ solar mass} = 1.99 \times 10^{30} \text{ kg}$$

Thus, $\Delta = \Delta_0 \times M/R^3$, and you can use the values of M and R directly from Data Table 1 to calculate the ratio in the second term of the above equation, and then multiply this factor by the Sun's density in order to obtain the star's average density for each of the selected stars.

13. For the main-sequence stars listed in **Data Table 3**, copy the values of **radius** and **mass** from **Data Table 1**, calculate the ratio M/R^3, and then calculate their average densities by multiplying this ratio by the Sun's density.

DATA TABLE 3

Star name	Radius, R (solar radii)	Mass, M (solar mass units)	Ratio M/R³	Average Density kg/m³
Bellatrix				
Spica				
Fomalhaut				
Altair				
Epsilon Eridani				

Question 6: How does the average density of a main-sequence star compare with the density of water, 1000 kg/m³?

D. Variable Stars

Section 21–5 in Freedman and Kaufmann, *Universe*, 7th Ed., discusses Cepheid and RR Lyrae stars.

Section 21–5 and Box 26–1, Tools of the Astronomer's Trade in Freedman and Kaufmann, *Universe*, 7th Ed., discuss the estimation of distance using Cepheid variables.

Section 26-4 and Figure 26-12 in Freedman and Kaufmann, *Universe*, 7th Ed., place the Cepheid distance method in context with other distance methods. This distance ladder is also the subject of separate chapter in the present book.

Many stars are found to vary in brightness periodically with a wide range of oscillation periods. Some of these stars vary in brightness with precise regularity while others are much less regular. This variation in brightness is accompanied by coincident changes in temperature and size. When placed upon the H–R diagram, these stars are found to occupy positions that are related to their period. For example, a range of stars with regular periods of between 1 and 100 days are known as **Cepheid variables** after the first of this type discovered, δ Cephei. They occupy a vertical band above the main sequence known as the **instability strip**, extending across a range of luminosities through the yellow giant and supergiant regions. A similar class, the **RR Lyrae stars** with more restricted periods of less than one day, occupies a small region with a narrow range of luminosity at the bottom of the instability strip. An important relationship has been established between the period of Cepheid variables and their luminosity. Thus, merely recognizing that a star is a Cepheid variable and measuring its period establishes its intrinsic brightness. When combined with a measurement of its apparent brightness, this information can be used to provide a measure of its distance from the Sun. Cepheid variables are sufficiently bright that they can be identified in nearby galaxies. This method of distance measurement is one crucial rung in the so-called **distance ladder**, whereby astronomers estimate the overall scale of the universe.

Other stars with much longer and somewhat variable periods, the long-period variables, LPV, occupy a region on the cooler side of the H–R diagram, again extending over a wide range of luminosities.

Data Table 4 contains a list of several different variable stars that can be found using *Starry Night*™ and their properties plotted on your H–R diagram to illustrate the positions of variable stars on this diagram.

DATA TABLE 4

Name of Star	Temperature (K)	Radius (solar radii)	Luminosity	Absolute Visual Magnitude M_v	Period (days)	Type of Variable
T Vulpeculae					4.4 days	Cepheid
Y Ophiuchi					17 days	Cepheid
T Monocerotis					27 days	Cepheid
RR Lyrae					0.57 days	RR Lyr
T Centauri					90 days	LPV
Mira					331 days	LPV
R Serpentis					257 days	LPV
R Hydrae					415 days	LPV

14. For each variable star in **Data Table 4**, first **Find** the star and then click **Show Info** in the contextual menu for the star.
15. From **Info/Other Data**, enter the values of **temperature, radius, luminosity,** and **absolute visual magnitude** into **Data Table 4.**

You can now plot these stars on your H–R diagram to locate the approximate position of the instability strip containing Cepheid and RR Lyrae stars and the Long Period Variable region. You will see that they are all well above the Main Sequence and in the giant and supergiant regions of the diagram.

E. Conclusions

In this project you have emulated professional astronomers in plotting an H–R diagram for a selected set of typical stars, but without having to go to the effort of measuring spectra, star brightnesses, or star distances in the real world. However, you have the satisfaction that you have been able to utilize the best available data set in the world for this type of investigation, the Hipparchos database incorporated in *Starry Night*™. You have located a few representative white dwarf stars and variable stars for comparison with other classes of stars on this important classification tool, the Hertzsprung–Russell diagram.

MEASURING THE SPEED OF LIGHT

19

The speed of light is a fundamental parameter in science. It is basic to the theory of relativity, which teaches us that this speed is also the absolute limit to speed in our universe. Nothing can travel at a speed that exceeds this value.

In his book *Dialogue Concerning the Two New Sciences,* published in 1638, Galileo proposes a straightforward experiment to determine whether light is instantaneous, as Aristotle proposed, or travels with some finite speed. In Galileo's experiment, two observers equipped with shuttered lanterns were located upon adjacent hilltops a known distance apart. The first observer would uncover his lantern and, upon seeing the light from this lantern, the second observer would remove the shutter from his lantern. The first observer would then attempt to measure the time between uncovering his lantern and seeing the light from his colleague's lantern. The extremely high speed of light doomed this experiment to failure because the response times of the observers would have been far too slow to produce a meaningful result, as Galileo correctly reasoned. His suggestion that a finite speed for light could perhaps be measured by repeating the experiment over greater distances, with the participants using telescopes to observe the light signal, was also unlikely to succeed. With the modern value for

> Section 5–1 of Freedman and Kaufmann, *Universe,* 7th Ed., discusses the methods of Galileo and Romer for measuring the speed of light.

the speed of light of 299,792,458 meters per second, participants in Galileo's proposed experiment, sending their signal back and forth from mountaintops 15 kilometers apart would need to be able to detect a time difference of about one-ten-thousandth of a second, a time period far shorter than the response time of a human being.

Nevertheless, barely 38 years later in 1676, the Danish astronomer Olaus Romer published evidence that light is not instantaneous. Just as Galileo predicted, Romer made this discovery by using a telescope to observe light signals from very distant events. Interestingly, these events that Romer observed were the regular eclipses of Io, one of the four moons of Jupiter that Galileo had discovered in January of 1610.

A. Observing Eclipse Events of Io

The orbits of the four Galilean moons around Jupiter, as well as that of Jupiter around the Sun, lie close to the ecliptic plane. This geometry leads to several phenomena that are observable from Earth as the Galilean moons orbit Jupiter. These phenomena include transits of the moons and their shadows across the face of Jupiter as well as occultation and eclipse events in which the moons pass behind Jupiter and into or out of its shadow.

These periodic phenomena provided early observers with an initial method for determining the orbital periods of the Galilean moons accurately. However, as we shall see, these events also provided light signals that Romer was able to use to demonstrate that light does not travel instantaneously. His results led to an estimate for the speed of light but his method is fraught with difficulty. It was only when Earth-based methods were developed almost 200 years later that accurate measurements of this important parameter were made. It is nevertheless interesting to repeat this historically significant experiment by observing some of the events that provided Romer with the clue that the speed of light was finite but very large.

Not all of the phenomena mentioned above are observable from Earth during every orbit of a Galilean moon, even when the effects of daylight and the horizon are discounted. The visibility and sequence of the phenomena depend upon the relative positions of the Sun, the Earth, and Jupiter. For example, as seen from Earth, when Jupiter is between conjunction and opposition, Io disappears into eclipse before it is occulted by the west limb of Jupiter but reappears at its full brightness at the East limb of Jupiter, having already emerged from eclipse while still occulted. The reverse is true when Jupiter is between opposition and conjunction. Jupiter occults Io before eclipsing it and Io reappears some distance from the east limb of Jupiter as it finally emerges from eclipse.

Observationally, eclipse events are the easiest to time accurately. For example, in an eclipse of Io as seen through a telescope from Earth, Io gradually dims over 1 to 2 minutes as this moon gradually enters Jupiter's shadow before finally disappearing into full eclipse. This disappearance event marks a fairly well defined endpoint. Similarly, when Io emerges from eclipse, it reappears at a fairly sharp point in time and then gradually brightens as it emerges from Jupiter's shadow.

In *Starry Night™*, eclipse events are depicted somewhat differently but still provide a sharp endpoint in time for the event. In this program, when the field of view is set to greater than about 2 minutes of arc, an eclipse event of Io is depicted as a sudden brightening or dimming (depending upon whether Io is emerging from or entering eclipse) at the point when approximately one-half of the satellite's disk is covered by Jupiter's shadow. Also, even when fully eclipsed, Io does not disappear from view but remains visible as a smaller and dimmer object.

In this section you will use *Starry Night™* to observe an eclipse of Io and practice a technique for determining the time, to the nearest second, at which the event occurs. Later, you will use these methods to make further observations of the light signals of eclipse events from varying distances in order to calculate the speed of light.

1. Launch *Starry Night™*.
2. Select **File/Preferences** and under **Cursor Tracking (HUD)** choose only **Name** and **Object type** from the **Show** list.
3. Select the view **Speed of Light-A** under **Go/Observing Projects/Speed of Light**.

This telescopic view from the equator of Earth is centered upon Jupiter. Also in the view are Jupiter's four Galilean moons, with Io labeled. Stars have been turned off. With no stars visible in the sky, this will produce a view that is somewhat unrealistic but it will allow us to view Jupiter and its moons without the confusion of a star background.

4. Use the **Hand Tool** to identify each of the four Galilean moons.
5. Watch Io carefully and start the flow of time forward. As Io approaches Jupiter's west limb, it suddenly dims in brightness or even disappears as it enters Jupiter's shadow. This is the beginning of an eclipse event.
6. Allow time to continue to flow forward and watch for Io's reappearance at the left (east) limb of Jupiter, which occurs at approximately 9:16:00 A.M. You will notice that Io emerges from behind the planet at its normal brightness. In other words, Io has already moved out of Jupiter's shadow by the time it reappears at the east limb of Jupiter.

TIP: To review a sequence, use the **Edit/Undo** [Ctrl + Z] command to undo time flow.

Question 1: Is Jupiter at, before, or after opposition on January 6, 2005?

7. Stop time flow and change the time and date in the control panel to **1:50:00 P.M.** on **August 6, 2005.**
8. Identify the four Galilean moons using the **Hand Tool.** Notice that, although very close to Jupiter's west limb, Io is not in eclipse.
9. Change the time step to 300× and watch Io carefully. Notice that, on this date, Jupiter occults Io before eclipsing it.
10. Stop the flow of time and change the time in the control panel to **4:05:00 P.M.**
11. Resume time flow forward and watch for Io to reappear at the left (east) limb of Jupiter. You will notice that when Io moves out from behind Jupiter, it is very dim or not visible at all; in other words, it reappears in eclipse.
12. Continue to watch Io until it suddenly brightens as it emerges from eclipse; then stop the flow of time.

Question 2: Is Jupiter at, before, or past opposition on August 6, 2005?

13. Select the view **Speed of Light-A** again under **Go/Observing Projects/Speed of Light.**

In this observing project, you will need to determine the time of eclipse events of Io to the nearest second. An easy way to make this measurement is explained in the following steps.

14. Click the **minute** field of the time in the control panel.
15. Repeatedly press (or press and hold) the + **key** on the keyboard to advance time in steps of 1 minute. Stop pressing the + **key** as soon as Io dims.
16. Click the field for **seconds** in the control panel and repeatedly press (or press and hold) the − **key** on the keyboard to reverse time in steps of 1 second until Io brightens again.
17. With the seconds still highlighted, use the + and − keys on the keyboard to advance and reverse time until you can identify the exact second at which Io becomes dim.

Question 3: At what time, to the nearest second, does Io enter into eclipse on January 6, 2005 as seen from the Earth?

B. The Speed of Light

With *Starry Night*™ we can observe the same eclipse of Io from two different locations in the Solar System separated by a known and relatively large distance. In this way, we are using Galileo's method to measure the delay in the arrival of light from a particular event, in this case an eclipse of Io, from two locations. We do this by first observing an eclipse from Earth and then by observing the same eclipse from a location about 4 AU above the Earth's surface in a direction toward Jupiter. The relevant distances to Io can be obtained from the Info pane. The difference in distances can then be combined with the observed time difference between the two observations of the same eclipse event to measure the speed of light. While impossible in Galileo's time, one can imagine this type of experiment being carried out with a spacecraft in modern times.

18. Record in **Data Table 1** your answer to question 3 above (i.e., the time at which Io enters eclipse on January 6, 2005, as seen from Earth).
19. Open the **Info** pane for Io and note the value of Io's **Distance from Observer** under the **Position in Space** layer. (If the **Info** pane does not show information relevant to Io, right-click with the cursor positioned over this moon and activate the **Show Info** command in the contextual menu.) Record this value in Data Table 1.
20. Select the view **Speed of Light-B** under **Go/Observing Projects/Speed of Light**.

Your viewing location is now at a position about 4 AU above the point on the Earth from which you observed the eclipse of Io in the previous sequence. Jupiter is centered in the main view window, with Io to the right of Jupiter while Ganymede is just below and to the right of the planet in this 11-arc-minute field of view. Note that Jupiter appears much larger from this closer distance.

21. Use the **Hand Tool** to identify Ganymede in this view. Io is already labeled.
22. In the **Info** pane, note the **Distance from Observer** of Io and record the value in Data Table 1. (If the **Info** pane refers to an object other than Io, you can correct this by activating **Show Info** in the contextual menu that appears upon right-clicking with the cursor positioned over Io.)

While this viewing location is approximately 4 AU above the Earth, it is not 4 AU closer to Jupiter. Nevertheless, it is sufficiently closer to the observer that a finite speed for light will produce a measurable difference in the time of arrival of the light signal of Io's eclipse at this location compared to the observation from the surface of the Earth.

Question 4:   If light travels at a finite speed, do you expect the time at which Io enters into eclipse as seen from the location in space relatively close to Io to be earlier or later than the time the eclipse is seen on Earth?

Question 5:   At what time (earlier, at the same time, or later) would you expect the light signal of Io entering eclipse to arrive at Earth compared to when it arrives at the location in space closer to Io if, as Aristotle argued, light travels instantaneously (i.e., with infinite speed)?

23. Using the technique described in steps 14 to 17 above, identify the time, to the nearest second, at which Io enters eclipse as seen from this viewing location in space. Record the time in Data Table 1.
24. In the row labeled Differences in Data Table 1, calculate the difference in distance from Io (in AU) as well as the difference in the time of the event (in seconds) as seen from the two locations.

DATA TABLE 1

Observation Location	Distance to Io (AU)	Time of Eclipse
From Earth		
From space		
Differences		

Consider the significance of the results of the calculations you made in the previous step. The difference in distance between the two viewing locations and Io represents the extra distance that the light signal from Io's eclipse must travel to reach the Earth. The time difference is the time required for the light signal to travel this extra distance. This essentially replicates Galileo's experiment except that in this instance the light signal travels only one way and across a much larger distance.

Because speed is defined as the change in distance divided by the change in time, you have sufficient data not only to determine that light travels at a finite speed, but also to calculate its value.

25. Divide the difference in distance (in AU) between the two viewing locations by the difference in time of the arrival of the signal at the two locations (in seconds) to determine the speed of light in units of AU/second.
26. Convert this speed to kilometers per second, using the fact that 1 AU = 1.496×10^{11} m.

Question 6:   What is your result for the speed of light in meters per second?
How does your result compare to the modern value of 299,792,458 meters per second?

C. Olaus Romer's Discovery of the Finite Speed of Light

Between 1671 and 1676, the Danish astronomer Olaus Romer was compiling observations of eclipse events of Io at the Paris Observatory with the aim of developing a table for predicting these events. The motivation for these observations was to determine whether they might be useful for timekeeping onboard ship, a necessary requirement for determining longitude in ocean navigation before the development of accurate chronometers.

Romer found that the period of time between eclipse events of Io was relatively consistent and predictable when Jupiter was near to opposition, when the Earth was closest to Jupiter. However, when he used this observed period of time to predict future eclipse events for Io, Romer discovered that his predictions became more and more inaccurate as the distance between the Earth and Jupiter increased in the months from opposition to conjunction. The farther Jupiter was from the Earth, the more the actual times of the events lagged behind predicted times. On the other hand, in the months from conjunction to opposition, as the distance between the Earth and Jupiter decreased, the time lag became smaller until the observed time of the event caught up once more to the predicted time, derived from measurements at Jupiter's opposition.

This led Romer to conclude that light traveled at a finite speed because the lag between predictions and the observed eclipse events increased as the distance between Earth and Jupiter increased. As the distance between Io and Earth closed again, the light signal had less distance to travel and so arrived at Earth progressively earlier, allowing predictions and observations to coincide once more.

In the following sequence, you will emulate Romer's observations to obtain an estimate for the speed of light.

27. Select the view **Romer** under **Go/Observing Projects/Speed of Light**.

The view shows a close-up of Jupiter as seen from Paris, France on June 8, 1675 at 9:53:40 A.M. Io, which is labeled in this simulation, appears very dim or invisible because it is still in eclipse at this time. Two other moons, Europa and Ganymede, appear to the right of Jupiter.

28. Manipulate the time controls to find the precise time at which Io emerges from eclipse.
29. Open the **Info** pane and find the distance to Io from Earth under **Position in Space/Distance from observer**.

30. Record the Date, Time, and Distance to Io in Data Table 2 below.
31. Select a time step of 30,000× and observe Io as it orbits Jupiter. Stop time flow when Io emerges from behind Jupiter again and find the exact time at which Io emerges from eclipse in this orbit. Record the Date, Time, and Distance to Io in Data Table 2.

DATA TABLE 2

Orbit	Date	Time (h:m:s)	Distance (AU)
0			
1			
Difference in times (eclipse period)			

32. From the data in Data Table 2, calculate the eclipse period of Io in hours, minutes, and seconds and enter this time interval in the **Difference in times** column.

These steps have established a suitable period for predicting the times when Io would reappear from Jupiter's shadow if the speed of light were infinite. You can now advance time forward by the equivalent time for 15 orbits of Io around Jupiter and determine whether, when viewed from a distant Earth with the a finite (but very large) speed of light, this moon is actually seen to reappear at this time or whether there is a measurable time lag.

33. Use your calculated eclipse period to predict when Io will emerge from eclipse 15 orbits further in the future. To do so, multiply the hours, minutes, and seconds of the eclipse period of Io by 15. (For example, if you calculated an eclipse period for Io of 40 hours, 20 minutes, and 10 seconds, then you would have values of 600 hours, 300 minutes, and 150 seconds. This is simply an example; your numbers will be different.) Enter these values in the **Reference Table 1** below.

Reference Table 1: Time interval for 15 Io orbits

Hours	Minutes	Seconds

You can now advance time by this predicted time interval for 15 orbits of Io around Jupiter.

34. Change the time step in the control panel to the number of **hours** in 15 Io eclipse periods and click the single step forward button **once**. Do the same for the number of **minutes** and the number of **seconds** in 15 eclipse periods.
35. Note the Date and Time displayed in the control panel in the **Predicted** column of Data Table 3 below.

You have now advanced the time to the instant when Io should be re-appearing from the shadow of Jupiter, if the speed of light were infinite. You will note however that Io still appears to be within this shadow.

36. Determine the actual time at which Io appears to emerge from eclipse during this orbit and record the date and the time in the Observed column of Data Table 3.
37. Calculate the time lag in seconds between the predicted time of Io's emergence from eclipse and the observed time of this event and record the result in the Time Lag column of Data Table 3.

DATA TABLE 3

Predicted		Observed	Time Lag (seconds)	Distance (AU)
Date	Time	Time		

38. From the **Info** pane, find the distance to Io from Earth at the point in time at which Io is emerging from eclipse on this orbit (July 6, 1675), and note this distance in Data Table 3.
39. Calculate the change in distance, in AU, between Io's distance on June 10, 1675, and its distance on July 6, 1675.

You have now demonstrated that light took a significant time to travel this differential distance between the positions of Jupiter. You can calculate the required speed of this light in order to have covered this distance in the measured time lag.

40. Divide the change in distance in AU by the Time Lag in seconds to obtain the speed of light in units of AU per second.

Having observed the time lag between eclipse events for Io when Jupiter is between opposition and conjunction, you can now observe eclipse events for Io when Jupiter is between conjunction and opposition.

41. Change the Date in the control panel to **December 25, 1675 A.D. (12/25/1675 A.D.)**.
42. Change the Time in the control panel to **7:47:30 A.M.**

In this view, Io is seen to be very close to the opposite limb of Jupiter from that in the previous sequence. Jupiter has just passed conjunction and Io is about to enter eclipse.

43. Highlight the **Seconds** field of the time and find the exact second at which Io enters eclipse (i.e., becomes dim in the view). Record the **Date** and **Time** for this event in the Orbit 0 row of Data Table 4 below.
44. Open the Info pane and record the distance to Io at the moment it enters eclipse on Orbit 0 in Data Table 4.
45. Change the time step to 30,000× and observe Io for one period, stopping time flow when it approaches Jupiter in its next orbit. Use the technique of steps 14 to 17 above to determine the exact time at which Io is eclipsed on this orbit. Record the Date, Time, and Distance to Io in Data Table 4 for this orbit (orbit 1).

DATA TABLE 4

Orbit	Date	Time (h:m:s)	Distance (AU)
0			
1			
Difference in times (eclipse period)			

46. Use your calculated eclipse period to predict when Io will enter eclipse 15 orbits further in the future. To do so, multiply the hours, minutes, and seconds of the eclipse period of Io by 15. Enter these values in the **Reference Table 2** below.

Reference Table 2: Time interval for 15 Io orbits

Hours	Minutes	Seconds

You can now advance time by this predicted time interval for 15 orbits of Io around Jupiter.

47. Change the time step in the control panel to the number of **hours** in 15 Io eclipse periods and click the single step forward button **once**. Do the same for the number of **minutes** and the number of **seconds** in 15 eclipse periods.
48. Note the Date and Time displayed in the control panel in the **Predicted** column of Data Table 5 below.

You have now advanced the time to when Io should be entering the shadow of Jupiter. You will note however that Io is already dim (i.e., in eclipse) and the actual event has occurred earlier than expected.

49. Determine the actual time at which Io enters eclipse during this orbit and record the Date and the Time in the Observed column of Data Table 5.
50. Calculate the time difference in seconds between the predicted time of Io's entrance into eclipse and the observed time of this event and record the result in the Time Lag column of Data Table 5.

DATA TABLE 5

Predicted		Observed	Time Lag (seconds)	Distance (AU)
Date	Time	Time		

You will note that the eclipse of Io appears to occur earlier than expected. This is because the distance between Jupiter and Earth is now decreasing and the light signal of this event requires less time to cover the reduced distance between the Earth and Io. Again, this suggested to Romer that light traveled at a finite speed.

You can now calculate the required speed of this light in order for it to have arrived this many seconds early because of the reduced distance.

51. From the **Info** pane, find the distance to Io from the Earth at the point at which Io is entering eclipse on this orbit (January 22, 1676), and note this distance in Data Table 5.
52. Calculate the change in distance, in AU, between Io and the Earth over the period between the second measured eclipse in December and the present eclipse.
53. Divide the change in distance in AU by the Time Lag in seconds to obtain the speed of light in units of AU per second.

You will notice that your two measurements for the speed of light are quite different. Romer made numerous such measurements of eclipse events of Io over several years in an attempt to derive a value for the speed of light. Because the distances involved were not well known at the time, Romer (or perhaps Huygens using Romer's data) estimated the speed of light to be about 2.1×10^8 m/s.

Later measurements have shown that, even if Romer had known the precise value of an Astronomical Unit in kilometers, he could not have produced a reliable value for the speed of light using this method. It is found (and using *Starry Night*™ reinforces this conclusion) that the measurement is very sensitive to the chosen initial period for predicting future eclipses. Even an inconsistency in the selected eclipse period of one-tenth of a second produces a significantly different result and Romer could not have measured reappearance times of Io to this precision.

Nevertheless, you can take the average of your two measurements to estimate the speed of light using Romer's method.

54. Calculate the average of your two measurements for the speed of light in AU/s. Convert this speed to m/s using the modern value for the distance of an Astronomical Unit of 1.496×10^{11}m.

It is now recognized that this measurement is very difficult. It was almost 200 years before reliable optomechanical laboratory methods were able to measure this speed correctly.

It is worth noting that one other early astronomical method for measuring the speed of light was developed in the interim. In 1680, Jean Picard of France had observed that the apparent positions of stars in the sky varied by about 20 arc seconds from their true position over a 1-year cycle, with all stars in a given part of the sky showing exactly the same effect. This phenomenon is called "the aberration of starlight" and is different from, and much larger than, the parallax effect that is smaller for stars at larger distances, an effect that was unobserved at that time. James Bradley, Astronomer Royal of England, explained aberration of starlight in 1729 as being caused by the effect of the velocity of the Earth upon the apparent direction of the light from the stars. A common analogy to this situation is the way in which, when rain is falling vertically, one points an umbrella at an angle to the vertical to shield oneself from the rain when one is walking forward. This observation in principle allows a determination of the speed of light because the faster the light moves, the smaller the angle of aberration for a given orbital speed of the Earth. Unfortunately, although the angle of aberration could be measured quite accurately, the size of the Astronomical Unit, and therefore the orbital speed of the Earth, was not well known in 1729. Bradley confirmed Romer's result that the speed of light was finite but very large, but, like Romer, he was unable to obtain an accurate value.

D. Conclusion

The two approaches in this project have demonstrated that light travels at a finite speed. Careful measurements allowed you to calculate a value for this very high speed. In the first section, you were able to measure the difference in times of appearance of Io being eclipsed by Jupiter from two positions a known distance apart and thereby derive an accurate value for the speed of light. In the second section, you were able to repeat Romer's historic measurements of the difference between predicted and observed eclipse times of Io when the Earth is at different distances from Jupiter. Careful measurements with Jupiter near opposition (closest to the Earth) and conjunction (furthest from the Earth) allowed you to calculate an approximate value for this speed, and demonstrate that the speed of light is finite but very large, just as Romer did more than 300 years ago.

THE DISTANCE LADDER 20

It has been less than 100 years since astronomers came to appreciate the immense size of our universe. As recently as 1920, the astronomer Harlow Shapley argued in a celebrated debate with Heber Curtis of Lick Observatory that the Milky Way encompassed the entire universe. In the period before this debate, Shapley had used the distribution of globular clusters in the sky to deduce the position of the Sun within the Milky Way and to provide a better insight into the size of this galaxy. However, he also concluded erroneously that "spiral nebulae" were within the Milky Way and argued this point forcefully in the debate. It was only three years later, in 1923, that Edwin Hubble, working at Mount Wilson Observatory, found very strong evidence to prove Shapley wrong. Hubble discovered that the "spiral nebula" known as the **Andromeda Nebula** was in fact a completely separate **galaxy** far removed from our own Milky Way. This initial discovery was soon followed by the identification of many other galaxies, or "separate worlds," to use Curtis's phrase. In the 80 or so years since Hubble's discovery, astronomers have found evidence that the universe contains billions of other galaxies, and have developed techniques for estimating their distances. In this way, astronomers have systematically probed the limits of the observable universe, discovering objects at distances so vast that it has taken billions of years for their light to reach us. Our understanding of the size and evolution of the universe has been profoundly altered by these discoveries.

Astronomers use various methods to measure distances in the cosmos. Each of these methods is useful over a particular range of distances. Fortunately, the ranges overlap to some extent, allowing astronomers to calibrate further-reaching techniques against more reliable closer-range methods. In this way, a so-called **distance ladder** has been assembled, each rung of the ladder representing a technique that extends the range of measurable distances in the universe.

> See Section 26-3 of Freedman and Kaufmann, *Universe*, 7th Ed., for a discussion of distance measurement in the universe, and Figure 26-12 for an illustration of the distance ladder.

In this project, you will "climb" the distance ladder. You will ascend each "rung" by making observations that use or simulate the distance measuring technique appropriate to that "rung."

A. Diurnal Parallax of the Moon

The first rung of the distance ladder is the **parallax** effect. This technique has been used to determine distances to objects that are up to approximately 500 parsecs, or about 1600 light-years, away.

Parallax is the apparent shift in position of a relatively nearby object against a more distant background, caused by a change in an observer's position. A simple experiment demonstrates the parallax effect. Close your right eye and note the position of your outstretched hand against a distant background when observed along the line of sight of your left eye.

Now close your left eye and open the right. The difference in the line of sight between your two eyes causes your hand to appear to "jump" to the left against the background. The closer an object, the more pronounced the effect. You can verify this by repeating the experiment with your hand held closer to your face.

The parallax effect also depends upon the distance between the two viewing points, the observational baseline. In the experiment above, the baseline is the distance between your two eyes (specifically, the pupil-to-pupil distance). If you try to repeat the experiment by observing, instead of your outstretched hand, an object much farther away, for example a tree trunk or signpost about 30 meters away, and compare its position against an even more distant background when viewed through each eye independently, it is unlikely that you will notice any parallax. However, if you note the position of the tree trunk or signpost against the background from one spot and then move 5 to 10 meters to the right or left, the parallax effect becomes obvious once more. Thus, increasing the observing baseline increases the sensitivity of the parallax effect and thereby increases its range of effectiveness.

This form of parallax is called **trigonometric parallax** and, since the distance to the object of interest is generally much greater than the baseline of observation, it is really an application of the small-angle formula.

The angle, θ, in radians, through which an object at a distance d shifts against a much more distant background when observed from two different locations a known distance, X, apart is given by the small-angle equation

$$\theta = \frac{X}{d} \qquad \text{(Equation 1)}$$

where X and d must be in the same units, and d must be much larger than X.

Parallax angles in astronomy are usually very small and usually expressed in arc seconds, whereas the formula above assumes that the angle is expressed in radians. For convenience, therefore, astronomers usually rewrite Equation 1 with a factor to convert radians to arc seconds. An angle of 2π radians is a full circle or 360° (thus, 1 radian = 57.3°), and a full circle is equal to $360 \times 60 \times 60$ arc seconds. Our conversion factor is then given by

$$2\pi \text{ radians} = 360 \times 60 \times 60 \text{ arc seconds}$$
$$\text{or} \quad 1 \text{ radian} = 206{,}265 \text{ arc seconds}$$

Rearranging Equation 1, the distance (in the units of the baseline, X) to an object in terms of its parallax angle in arc seconds becomes

$$d = \frac{206{,}265\, X}{\theta} \qquad \text{(Equation 1)}$$

Equation 1 shows that, for a given baseline distance X, the larger the distance, d, to the object, the smaller the angle of parallax. Similarly, for any given distance, d, increasing the baseline produces a larger and more obvious angle of displacement. You saw this in the experiment above when you were unable to detect a parallax shift of a distant tree or signpost when using the baseline of the distance between your eyes but had no problem seeing the displacement when you made observations from locations 5 or 10 meters apart. Because very small angles of displacement are difficult to detect, and measurement precision is often limited by atmospheric seeing, astronomers want the largest baselines available for making parallax measurements. Earthbound astronomers have two potential baselines at their disposal: the diameter of the Earth, and the diameter of the Earth's orbit around the Sun.

Parallax from a baseline limited to the diameter of the Earth is called **diurnal** or **geocentric parallax**. The strict definition of diurnal parallax is the angular displacement of an object when observed from a location on the surface of the Earth and a hypothetical observation made from the center of the Earth. In practice, diurnal parallax measurements are made by observing an object from a single location at two different times, taking advantage of the fact that the Earth's rotation carries the observing location through a known distance between observations. Because of its relatively short baseline, the range of diurnal parallax is limited to measuring distances within the solar system.

Question 1: To obtain the maximum possible baseline from which to measure diurnal parallax, where would you place the observing location and how far apart in time would you make your observations?

Question 2: What does your answer to question 1 imply about when these observations should be made with respect to the rising and setting times of the object whose parallax is being measured?

Question 3: Assuming that the smallest displacement angle that can be measured is 1 arc second, and that the baseline is equal to the diameter of the Earth (12,756 km), what is the maximum distance, in kilometers, for which an object will reveal a measurable parallax?

Question 4: If 1 Astronomical Unit (AU) = 1.496×10^8 km, what is the distance in Question 3 in AU? (1 AU is the average distance between the Earth and the Sun.)

A disadvantage of diurnal parallax is that any independent motion of the observed object that occurs during the time between observations must be taken into account unless two coordinated observations from two independent sites are made. An alternative that avoids this problem, and which *Starry Night*™ can simulate, is to observe the object from two different locations at the same point in time. To achieve the maximum range, the two locations should be at the equator and separated by 180° of longitude, thereby producing a potential baseline equal to the equatorial diameter of the Earth. The baseline should be perpendicular to a line drawn from the center of the Earth to the object under observation. This condition will be fulfilled if the two locations are 90° in longitude on either side of the longitude on Earth where the object is on the meridian, defined as the object being in transit, at the time of the observations.

In the next steps, you will apply these concepts to the measurement of the parallax of the Moon by observing the Moon from two equatorial locations on opposite sides of the Earth.

1. Launch *Starry Night*™.
2. Select **File/Preferences...** from the menu and under **Cursor Tracking (HUD)** in the Preferences dialog choose only **Name, Object type,** and **Apparent Magnitude** from the **Show** list.
3. In **File/Preferences...**, choose **Brightness/Contrast** from the drop-down list and increase **Star brightness** to maximum.
4. Select the view **Moon-90E** under **Go/Observing Projects/Distance Ladder.**

The view shows the waning crescent Moon as seen from a location at 90° E on the equator of Earth at 3:08:05 A.M. on November 9, 2004. The star TYC281-1009-1 is labeled as a convenient reference star against which to measure the parallax of the Moon.

5. Measure the angular separation from the Moon to the star TYC281-1009-1. Record this result in Data Table 1 below.
6. Select the view **Moon-90W** under **Go/Observing Projects/Distance Ladder.**

The view is of the same region of sky but as seen from the equator of Earth at longitude 90° W.

7. With the view showing the Moon and the reference star as seen from longitude 90° W, measure the angular separation between the Moon and the reference star TYC281-1009-1 from this location. Record this measurement in Data Table 1 below.
8. In the Edit menu, click **Undo "Go" Item** and **Redo "Go" Item** alternately to select the views **Moon-90E** and **Moon-90W** and see the Moon switch between two apparent positions because of parallax.

DATA TABLE 1

Longitude	Angular Separation from TYC281-1009-1 (° ' ")
90° W	
90° E	
Difference	

9. Calculate the displacement angle of the Moon produced by the parallax effect by subtracting the smaller value in the data table from the larger value (in degrees, arc minutes, and arc seconds).
10. Convert the displacement angle calculated in the previous step to arc seconds. (Multiply the degrees by 60, add to the minutes, multiply the total by 60, and add to the seconds).
11. Using the small-angle formula, d (km) $= 206,265 \times 12,756$ (km)$/\theta$ (in arc seconds), calculate the Moon's distance d, in kilometers.

Question 5:   What is the distance to the Moon as calculated from its parallax?

Question 6:   Using your observations above and your understanding of the small angle formula, without doing any further calculations, what is the angular size of the Earth, as seen from the Moon on November 9, 2004 (in degrees, arc minutes, and arc seconds)?

12. To check the accuracy of your answer to question 4, position the cursor over the Moon in the view, open the contextual menu and choose **Show Info**. In the **Info** pane, expand the **Position in Space** layer and check the **Distance from Observer** to obtain the distance to the Moon.
13. Close the Info pane.

The time and locations of the observations you made in the previous steps were carefully chosen so that the baseline was perpendicular to the line joining the centers of the Earth and the Moon, thereby ensuring that the baseline distance was equal to the diameter of the Earth. To satisfy this condition, the two positions MUST be 90° either side of the longitude on the Earth where the Moon is crossing the meridian, that is, where the Moon is in transit. This geometry is shown in the diagram below, with A and B the observing positions, equally spaced around the point T where the Moon is in transit.

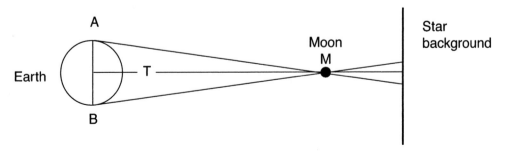

14. Select the view **Moon-0** under **Go/Observing Projects/Distance Ladder.**
15. Note the time in the control panel.

The view is of the same region of sky, but it is now as seen from longitude 0°E on the Earth's equator. The time is also the same, although again the local time is different because of the different longitude.

16. Open the contextual menu over the Moon and choose **Show Info.** Under the **General** layer, note the time of transit of the Moon from this location on the equator of Earth, at 0° longitude.
17. Close the Info pane.
18. You can demonstrate that the Moon is on the meridian (i.e., in transit) at this time from this location by clicking on and off the **Local Meridian** in the **View Options/Guides** pane.

B. Diurnal Parallax of the Asteroid Vesta

A photograph of Vesta by the Hubble Space Telescope is shown in Figure 17–5 of Freedman and Kaufmann, *Universe,* 7th Ed.

In this section, you will follow the principles outlined above to measure the diurnal parallax of a more distant solar system object, the asteroid Vesta. It is useful to translate distances from kilometers to Astronomical Units (AU), where 1 AU = 1.496×10^8 km.

Thus, the small-angle formula

$$d \text{ (km)} = \frac{(206{,}265 \times 12{,}756)}{\theta \text{ (arc seconds)}}$$

becomes

$$d \text{ (AU)} = \frac{(206{,}265 \times 12{,}756)}{(1.496 \times 10^8 \times \theta \text{ (arc seconds)})}$$

or

$$d = \frac{17.59}{\theta}$$

where d is in AU and θ is in arc seconds.

19. Select the view **Vesta-130E** under **Go/Observing Projects/Distance Ladder.**

The view is from the equator of Earth from a longitude of 130° E in Atlas mode, which renders the Earth transparent to avoid the interference of the horizon with some observations. This view shows the asteroid Vesta between two reference stars, TYC 1883-1567-1 and TYC1883-1884-1, in a field of view 6 arc minutes wide. The date is February 18, 2006. The time, 6:01 P.M., is in Universal Time (UT), although this is not stated on the screen. This time will be the same in the next view shown from a different location, 180° away in longitude.

20. Select the view **Vesta-50W** under **Go/Observing Projects/Distance Ladder.**

The view is from the equator of Earth at 50° W, at the point diametrically across the Earth from the previous view, at the same date and time (also given in UT). You will note that the reference stars have remained stationary, whereas Vesta appears to have "jumped" slightly as a result of parallax.

21. In the **Edit** menu, click **Undo "Go" Item** and **Redo "Go" Item** alternately to select the views **Vesta-130E** and **Vesta-50W** and watch Vesta switch between the two apparent positions because of parallax.

You can now measure this apparent motion using the celestial coordinate system as a reference frame. The stars remain fixed against this reference frame, as you can see by switching between the two views 180° apart.

22. To measure the diurnal parallax of Vesta and calculate its distance, return to the view **Vesta-130E.**
23. Use the **Hand Tool** to measure the spacing between Vesta and a convenient vertical line of Right Ascension.
24. Return to the view **Vesta-50W.**
25. Use the **Hand Tool** to measure the spacing between Vesta and the same line of Right Ascension.
26. Calculate the difference in these measured spacings to determine the observed shift in the position of Vesta as a result of the movement of the observing positions.
27. Insert your calculated difference into the formula above ($d = 17.59/\theta$) to calculate the distance to Vesta in AU.
28. Open the contextual menu over Vesta and select **Show Info.** Compare your result for the distance to Vesta with the value for **Distance from Observer** under the **Position in Space** layer in the **Info** pane.

Your result for Vesta is probably less accurate than your result for the Moon. Because as the distance to the object increases, the parallax shift decreases proportionally and the uncertainties in the measurement become more significant. (In *Starry Night*™, the precision of angular measurements for angles greater than 10 arc seconds is limited to 1 arc second and places a further limit on the accuracy achieved with this method.) Diurnal parallax thus has a very limited range and is inadequate for measuring distances much larger than several AU.

C. Parallax: Stellar Objects

Stellar parallax is discussed in Section 19-1 of Freedman and Kaufmann, *Universe*, 7th Ed.

Measuring the parallax of more distant objects, such as stars, requires a longer baseline than the Earth's diameter. The baseline used to determine stellar distances is the diameter of the Earth's orbit: The angular shift of a nearby star is measured against the background of more distant stars when observed from the Earth at two times that are separated by 6 months, during which time the Earth has moved to the opposite position in its orbit about the Sun, a distance of 2 AU. This method is called **stellar** or **heliocentric** parallax.

When measuring stellar parallax, astronomers also use a different unit of distance related specifically to parallax, the **parsec**. The **angle of parallax**, p, in arc seconds, is defined as the parallax shift when the baseline is 1 AU (the radius, or half the diameter, of the Earth's orbit). Thus p is equal to one-half of the measured parallax shift in arc seconds when the measurements are made 6 months apart.

With p defined in this fashion, the distance to the star in parsecs is found from the formula

$$d = \frac{1}{p}$$

The reason for using one-half of the value of the measured angular shift is to avoid the factor of two that would otherwise arise. Seen from another perspective, a parsec is equivalent to the distance from which the average radius of the Earth's orbit, 1 AU, would span an angular size of one arc second.

Despite the larger baseline used in stellar parallax observations, even the nearest stars have parallax angles of less than 1 arc second. Stellar parallax is thus a very subtle and difficult measurement to make. It was not until 1838 that Wilhelm Friederich Bessel made the first successful measurement of stellar parallax. In recent years, astronomers have been able to use modern ground-based telescopes to measure parallax angles of 0.01 arc second. The most recent and major advance in this important field of astrometry has been the satellite Hipparcos, whose name is an acronym for **High Precision Parallax Collecting Satellite**. Despite being placed in a wrong orbit, this satellite revolutionized our knowledge of star distances and positions by measuring the parallaxes of more than 100,000 stars to an accuracy of 0.001 arc second in the early 1990s. Star positions and properties derived from this space mission are included in the Hipparcos database and its companion, the Tycho database, and these databases have been incorporated into *Starry Night*™, hence the HIP and TYC prefixes associated with stars in this program.

> **Question 7:** Assuming that a parallax angle (i.e., one-half of the displacement angle) of 0.01 arc second can be reliably measured, what is the maximum distance (in parsecs) that can be determined from parallax observations?
>
> **Question 8:** What is the range of parallax measurements for the satellite Hipparcos in parsecs?

Starry Night™ cannot measure parallax angles with sub-arc second resolution. Even for the bright and close star, Alpha Centauri, for example, the parallax motion is not seen when observed at 6-month intervals although its simulated **proper motion** across the sky is observed. Therefore, we will use a different method for quantitative measurements of stellar parallax with *Starry Night*™. In the next sequence, imagine yourself attached to the end of a very long pole that is fastened to, and rotates with, the Earth. As the Earth rotates through 180° in 12 hours, the location of the end of the "flagpole" changes by a linear distance equal to the diameter of the Earth plus twice the length of the pole, thus generating whatever baseline we need. Through this artificial expansion of the baseline, you will be able to use *Starry Night*™ to demonstrate stellar parallax.

29. Select the file **Flagpole** under **Go/Observing Projects/Distance Ladder.**

The view is of a wide star field (100°-wide FOV). The celestial equator (horizontal line) and meridian (vertical line) intersect near the center of the view. Your viewing location is at the end of a "flagpole" that is 1.325 light-years long and planted in the Earth at longitude 0° at the equator.

30. Run **Time Forward.**

N.B.: The small diameter of the Earth can be neglected compared to the length of the flagpole.

As the Earth rotates, it carries your position in a circular orbit 2.65 light-years in diameter around the center of the Earth with a period of 24 hours. Apart from the fact that this "orbit" is centered on the Earth, the result of running Time forward is to produce a parallax effect similar to but much larger than that which astronomers observe from the surface as the Earth moves through its orbit around the Sun.

If you look closely at the motion of the stars in the field, you will note that some move a lot while others move relatively little. The nearer the star, the larger is its apparent motion. Stars positioned on the celestial equator appear to move in straight lines back and forth while stars above and below your "orbital" plane move in clockwise and counterclockwise ellipses, respectively.

31. With time continuing to flow, drag the view window downward so that you are looking at the northern sky. You will notice that the pattern of apparent motion of the stars becomes more nearly circular as the view direction gets progressively closer to the North Celestial Pole.
32. Drag the view upward so that the gaze direction is toward the South Celestial Pole. Notice that the sense of rotation has switched to counterclockwise. With this southern pole centered in the main view window, the very large circular motion of the close star **Rigil Kentaurus** is very obvious.
33. Stop the time flow and identify this star with the **Hand Tool**. Open the contextual menu for the star **Rigil Kentaurus** and choose **Show Info** to display the **Info** pane. Note its **Distance from observer** under the **Position in Space** layer.

There is one significant difference between this simulation and the real parallax motion of stars caused by the Earth's orbital motion. In the simulation, the Celestial Poles are the points around which the motion will be circular, whereas in reality it is the Ecliptic Poles that are the centers of real parallax motion.

This same flagpole technique will be used in the next sequence of observations except that the pole is 5000 AU long and establishes a baseline of 10,000 AU. This baseline is sufficient to make quantitative estimates of the distance to several stars by measuring the resulting parallax effect.

We saw above that, for an orbital radius of 1 AU, the distance, d, in parsecs, is related to the parallax angle, p, by $d = 1/p$. Thus, $p = 1/d$. We also saw in Section A that, for an object at a given distance, the parallax angle is proportional to the length of the baseline. Then for an orbital radius of 5000 AU, the parallax angle will be $p = 5000(1/d) = 5000/d$, giving

$$d \text{ (pc)} = \frac{5000}{p \text{ (arc second)}}$$

where p is half the observed angle of displacement when the baseline is the diameter of the orbit.

DATA TABLE 3

Star	Displacement Angle (")	Parallax Angle (")	Calculated Distance (pc)	Actual Distance (pc)
Altair				
Alshain				
Xi Aquilae				
Tarazed				

For each star listed in Data Table 3 above, complete the following steps.

> **IMPORTANT:** If you accidentally drag the view while making this measurement, use **Edit/Undo Scroll** (or the keyboard shortcut Ctrl + Z) and try the measurement again.

34. Select the appropriately named view from **Go/Observing Projects/Distance Ladder**.
35. Drag the view to the left so that the labeled star is bisected by the left edge of the window.
36. Click the **Single Step Forward** button in the control panel **once** to advance time by 12 hours. This moves your viewing location 180° around the Earth to a point 10,000 AU from the original position. (Note that the Fields of View used to display each star differ. Consequently the **apparent** motion of the star on the **screen** is not directly related to the measurement of the parallax angle in arc seconds.)
37. Measure the angular separation between the star and the left edge of the window (the star's previous position), keeping the line that extends from the star horizontal.
38. Convert the angle to arc seconds by multiplying the number of arc minutes in the measurement by 60 and adding the result to the number of arc seconds in the measurement (arc minutes × 60 + arc seconds). Record the displacement in arc seconds in Data Table 3.
39. Divide the displacement angle by two to obtain the parallax angle in arc seconds.
40. Calculate the distance to these stars in parsecs from their displacement angles using the formula above, $d = 5000/p$, where p is in arc seconds.
41. Open the Info pane and note the distance to the star under the **Position in Space/Distance from observer**. Divide this distance in light-years by the factor 3.26 to convert the distance to parsecs (1 parsec × 3.26 light-years) and enter the result under the column **Actual Distance** in Data Table 3.

D. Spectroscopic Parallax

The next rung of the cosmic distance ladder is called **spectroscopic parallax,** and has a range of up to 10 kiloparsecs (about 32,000 light-years). Despite its name, this method for determining astronomical distances does not use the measurement of parallax angles. Rather, spectroscopic parallax relies on the concept of **standard candles,** objects whose luminosity or absolute brightness is known. If we know the absolute brightness of a particular star and compare this to its apparent brightness as seen from the Earth, we can determine the star's distance from Earth using the inverse square law.

The method of spectroscopic parallax thus depends upon astronomers identifying "standard candles" and determining their absolute brightness. One such method utilizes the spectrum of a star to classify it and place it upon the Hertzsprung-Russell (H-R) diagram, the widely used summary diagram for stellar astronomy. This will establish the star's luminosity. By comparing the apparent brightness of the star to its actual luminosity as determined from the H-R diagram, the inverse square law can be used to calculate the distance to the star.

> Section 19-8 in Freedman and Kaufmann, *Universe*, 7th Ed., discusses the principles of spectroscopic parallax.

The next sequence of observations uses this method to determine the distances to the stars in Data Table 4 below.

DATA TABLE 4

Star	App. mag. (m)	Abs. mag. (M)	Distance (pc)
Altair			
Alshain			
Xi Aquilae			
Tarazed			

For each of the stars listed in Data Table 4 above, complete the following steps.

42. Open the appropriately named view under **Go/Observing Projects/Distance Ladder.**
43. Open the **Info** pane, and under the **Other Data** layer find the **Apparent magnitude** and **Absolute magnitude** of the star. Record these values in the appropriate columns of Data Table 4 above.

The absolute, or total, energy output of these stars has been established by careful observations of their spectra and other properties to determine their spectral-luminosity classification, in order to place them on an H-R diagram. For example, Altair is a main-sequence A7 star. From its position on the H-R diagram, this star has an absolute magnitude of 2.19.

The difference, $m - M$, between its apparent magnitude as seen from the Earth, m, and its absolute magnitude, M ($m - M$) is called the **distance modulus** of the star. The distance modulus is related to the distance, d, to the star by the formula:

$$m - M = 5 \log d - 5$$

See Box 19–3, Tools of the Astronomer's Trade in Freedman and Kaufmann, *Universe*, 7th Ed., for a discussion of the distance-magnitude relation.

The apparent magnitude of the star is measured through observation. By rearranging the distance modulus formula, the distance to the star (in parsecs) can be calculated as follows:

$$d = 10^{[(m - M + 5)/5]}$$

44. Use the formula above to determine the distances to the stars listed in Data Table 4, and compare your results with those you obtained in Data Table 3. (Take care with the sign of the magnitude values.)

E. Population I Cepheid Variables

Variable stars are discussed in Section 21–5 of Freedman and Kaufmann, *Universe*, 7th Ed.

Cepheid variable stars are stars that have evolved away from the main sequence, and occupy the so-called **instability strip** on the H-R diagram. Cepheids vary regularly in brightness in a specific and identifiable pattern and get their name from the prototype for this class of stars, δ (delta) Cephei.

There are two features of Cepheid variable stars that make them extremely useful for measuring distances in the universe. First, Cepheids are intrinsically very bright, with luminosities ranging to more than 10,000 times the luminosity of the Sun. Consequently, they can be seen from distances of millions of parsecs. Second, a relationship exists between the period over which the brightness of a

Figure 21–17 in Freedman and Kaufmann, *Universe*, 7th Ed., shows the period-luminosity relationship for Type I and Type II Cepheids.

Cepheid varies and its average intrinsic luminosity. A measurement of this period is thus sufficient to determine the luminosity for this star. The astronomer Henrietta Leavitt established this relationship in 1912. In fact, there are now two different types of Cepheid variable stars known: Population I and Population II Cepheids. Each shows a specific period-luminosity relationship. Thus, in addition to identifying a star as a Cepheid, astronomers must also determine in which population the star belongs before a definitive luminosity can be assigned to it.

Another class of variable stars, the population II RR Lyrae stars, also show a period-luminosity relationship. RR Lyrae stars are generally not intrinsically as luminous as Population I Cepheid variables, and have a very narrow but well-defined range of luminosities. While they do not provide as great a range of distance measurement as Cepheids, they are much more common and have therefore been important in the verification of other rungs in the cosmic distance ladder. Also, in 1920, Harlow Shapley used the period-luminosity relationship of RR Lyrae stars in globular clusters to determine the distances to these star groups in our Milky Way Galaxy. From the distances and distribution of these globular clusters he was able to deduce the size and extent of the Milky Way Galaxy.

Using trigonometric and spectroscopic parallax, astronomers have been able to calibrate the period-luminosity relationship so that Cepheid variables can be used as standard candles. To measure the distance to a very distant Cepheid, the astronomer measures the period of its variation and, from the period-luminosity relationship, determines its absolute magnitude. Comparing the star's absolute magnitude to its measured apparent magnitude gives a distance modulus from which the star's distance can be calculated in a manner equivalent to that of spectroscopic parallax.

See Box 26–1, Tools of the Astronomer's Trade in Freedman and Kaufmann, *Universe*, 7th Ed., for a discussion of how Cepheid variable stars and supernovae are used to determine distances in the Universe.

Because they are intrinsically very luminous, Cepheid variable stars can even be resolved in other galaxies. In 1923, Edwin Hubble used the 100-inch telescope at Mount Wilson to observe Cepheid variables in what was then called the Andromeda nebula and, from the period-luminosity relationship, determined that this "nebula" was in fact an independent star system approximately 2 million light-years distant from the Sun. Since Hubble's time, Population I Cepheid variables have extended the distance ladder to a range of up to 30 Mpc (almost 100 million light-years).

DATA TABLE 5

Star	Actual Distance (ly)	Apparent Magnitude (m)	Period (days)	Estimated Distance (pc)	Estimated Distance (ly)
T Monocerotis			27.03		
Polaris			3.97		
Eta Aquilae			7.18		
DT Cygni			2.5		

For each of the stars listed in Data Table 5, above, complete the following steps.

45. Select the appropriately named view under **Go/Observing Projects/Distance Ladder**. (*N.B.*: In the case of **T Monocerotis** and **DT Cygni**, the star positions are shown but the star is too faint to be displayed in this version of *Starry Night*™. Nevertheless, the following steps can still be carried out.)
46. Place the cursor over the star's position and right-click to show the contextual menu. Activate **Show Info** to open the Info pane for the star and, from the **Position in Space** layer, obtain the actual distance to the star in light-years and record this in Data Table 5 above.

47. Under the **Other Data** layer of the Info pane (or from the HUD) find the apparent magnitude of the star and record this in Data Table 5.
48. The graph in Figure 1 below shows the period-luminosity relationship for Population I Cepheid variables. Notice that the scale for the period of the star, the X-axis, is logarithmic and begins with the value 1.0. Use the graph in Figure 1 and the period for the star that is already entered into Data Table 5 to estimate the absolute magnitude of the Cepheid variable you are observing.
49. Use the formula $d = 10^{[(m - M + 5)/5]}$ that you used in the previous section on spectroscopic parallax to calculate the estimated distance to the star in parsecs. Record the result of your calculation in the column labeled **Estimated Distance (pc)** in Data Table 5.
50. Convert the estimated distance to light-years by multiplying the result of the previous step by the factor of **3.26** and compare your estimates to the "actual" values that you recorded from the **Position in Space** layer of the Info pane in *Starry Night™*.

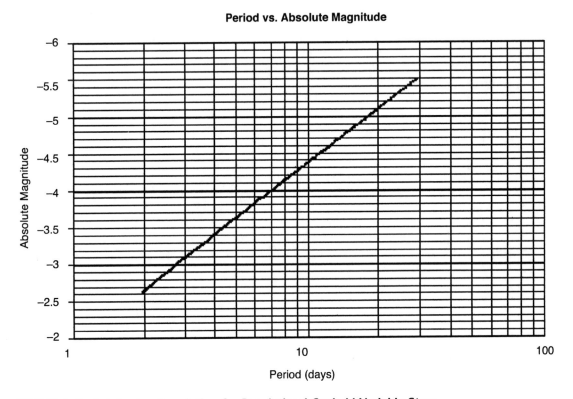

Period vs. Absolute Magnitude

FIGURE 1. Period-luminosity relation for Population I Cepheid Variable Stars

F. Type Ia Supernovae

Type Ia supernovae are discussed in Sections 22-6 and 22-7 of Freedman and Kaufmann, *Universe*, 7th Ed.

Type Ia supernovae occur when a white dwarf star in a binary system accretes sufficient matter from its companion star to blow itself apart in a massive thermonuclear explosion.

The maximum luminosity is very nearly the same for all Type Ia supernovae, a fact that has been established using Type Ia supernovae in relatively nearby galaxies whose distances are known from other standard candles. Because this maximum luminosity is known, Type Ia supernovae can themselves be used as standard candles. In fact, with luminosities of up to 3×10^9 times that of the Sun, Type Ia supernovae can be brighter than an entire galaxy and are the most luminous standard candles available to astronomers. Observations of these supernovae have extended the range of the distance ladder to approximately 1 billion parsecs (over 3 billion light-years).

G. The Tully-Fisher Relation

> The Tully-Fisher relation is discussed in Section 26-4 of Freedman and Kaufmann, *Universe*, 7th Ed.

With a range of up to about 100 Mpc, the Tully Fisher relation extends from the Cepheid variable rung and overlaps a considerable portion of the range of the Type Ia supernovae rung. The Tully-Fisher relation does not use standard candles. Rather, it relates the luminosity of a distant spiral galaxy to the width of the 21-cm radio emission line of hydrogen.

H. Masers

> The MASER method of distance measurement is independent of other methods. It is discussed in Section 26-4 of Freedman and Kaufmann, *Universe*, 7th Ed.

The recent discovery of a relation that allows distances to galaxies to be calculated from observations of masers is important in calibrating the rungs in the distance ladder because of its independence from other methods of distance determination. The term maser is an acronym for Microwave Amplification by Stimulated Emission of Radiation. Masers occur in vast molecular clouds within galaxies. Water molecules within these clouds are stimulated by the radiation of nearby stars to emit intensely in the microwave band of the electromagnetic spectrum. By measuring the Doppler shift of this radiation to determine the actual speed with which the molecular cloud is approaching or receding from Earth and comparing this with the apparent change in position of other masers in the galaxy that are moving across the line of sight, the distance to the galaxy can be calculated.

I. The Hubble Law

> The Hubble Law is discussed in Section 26-5 and in Box 26-2, Tools of the Astronomer's Trade in Freedman and Kaufmann, *Universe*, 7th Ed.

The final rung in the distance ladder is the Hubble Law. A few years after uncovering the true nature and distance to the Andromeda Galaxy, Edwin Hubble discovered a relationship between the distance to a galaxy and the apparent speed at which it is receding, as measured by the redshift of the lines in its spectrum. This relationship, called the Hubble law, has allowed astronomers to estimate distances to the limits of the observable universe. Furthermore, this relationship has profoundly altered our knowledge of the evolution of the universe.

A significant feature of Hubble's results is that all galaxies in the universe appear to be receding from each other. Cosmologists, supported by Einstein's General Theory of Relativity, have been able to conclude that this redshift is a consequence of the expansion of space itself. By extrapolating this expansion into the past, cosmologists have suggested that the universe was born 13 to 15 billion years ago from the explosion of an infinitesimal point of infinite density. Supporting evidence for this Big Bang cosmology has come from the study of the residual radiation of this explosion, the cosmic microwave background radiation.

The Hubble Law states that the distance to a galaxy, d, is proportional to its velocity of recession, v, and is expressed simply in the formula

$$d = \frac{v}{H_0}$$

where H_0 is a constant called the Hubble constant and is expressed in units of kilometers per second per megaparsec. The exact value of the Hubble constant is uncertain, within a range of 50 to 100 km/s/Mpc. Astronomers continue to use other rungs of the distance ladder, particularly Type Ia supernovae and the Tully-Fisher relation, to calibrate this important cosmological constant.

51. Exit *Starry Night*™ without saving changes to files.
52. Launch *Deep Space Explorer*™.
53. Once *Deep Space Explorer*™ has loaded, select **Edit/Find** from the menu and enter the name **NGC 2484** in the box labeled **Name contains** in the **Find** dialog window; then click the **Find** button.

Deep Space Explorer™ pans the view to bring you to within 109,000 light-years of the lenticular S0 galaxy catalogued as NGC 2484. Click and hold the left mouse button to drag the view in the screen in various directions in order to see the shape and size of this galaxy. The measured redshift, z, for this galaxy is equal to 0.042836. Use this to answer the following questions.

Question 9:   From the formula $z = v/c$, where z is the redshift, v is the recessional velocity and c is the speed of light, what is the recessional velocity of NGC 2484?

Question 10:   At the bottom of the main view window of *Deep Space Explorer*™, the distance to the Sun from the current viewing location in space is given. What is the distance to NGC 2484 in Megaparsecs?

Question 11:   Rearrange the equation $d = v/H_0$ to solve for H_0. Using your result from question 9 for the value of v, and the distance to NGC 2484 indicated on the bottom of the *Deep Space Explorer*™ main view window, what value does the *Deep Space Explorer*™ program use for the Hubble constant?

J. Conclusions

In this project, you have had the opportunity to review the techniques that astronomers employ to measure distances to objects in the cosmos. You had the opportunity to employ the gist of these techniques in software simulations. Finally, you learned about the distance ladder and how its various rungs not only extend the range of measurable distances but also verify and calibrate other rungs.

THE MILKY WAY 21

One feature of the night sky that must have attracted the attention of the very first "astronomers" is the ribbon of light that encircles it, the Milky Way. Sadly, the light pollution of today's modern world robs many of us of the opportunity to see this beautiful, subtle light that originates from within our home galaxy in the cosmos.

Galileo made the first telescopic observations of the Milky Way in 1609. His telescope resolved the diffuse band of light into countless individual stars. Later telescopic observers,

> The Milky Way Galaxy is discussed in Chapter 25 of Freedman and Kaufmann, *Universe*, 7th Ed. Chapter 26 discusses galaxies in general.

such as Charles Messier, began to discover other objects in the sky that appeared fuzzy in the eyepiece and could not be resolved into stars. Little was known of these mysterious objects except that some were eventually resolved into clusters of stars, while the rest remained nebulous. In the past century, modern observations of these objects have helped astronomers to:

1) Establish accurate estimates of the physical size of the Galaxy

2) Develop an understanding of its physical and dynamic structure

3) Determine that our Milky Way Galaxy is but one of billions of such galaxies in the universe

4) Construct theories about the origin and structure of the universe, based on the observed filamentary distribution and relative motion of the billions of galaxies within it

In this project, you will use *Starry Night™* to make observations of the Milky Way and some of these nebulous telescopic objects.

A. The Milky Way and the Broad Structure of the Galaxy

The term "Milky Way" refers, in its strictest sense, to the appearance of the Galaxy as seen from Earth, specifically to the diffuse ribbon of luminosity that winds its way among the constellations and encircles our sky.

> 1. Launch *Starry Night™*.
> 2. Select the view **Milky Way-A** under **Go/Observing Projects/Milky Way.**

> **TIP:** You can adjust the brightness of the Milky Way in this view. Open the **View Options** pane, expand the **Stars** layer, then place the cursor over **Milky Way** and click the **Milky Way Options** button that appears. The **Brightness** slide bar allows you to adjust the Milky Way brightness.

The view is of the southern sky from Cornwall, Canada. The time is midnight on the date of the summer solstice. The Teapot and Fish Hook asterisms are visible near the southern horizon and the band of the Milky Way sweeps diagonally upward to the left.

Were you to go outside at the time and place of this simulation and look south at the teapot asterism, you would see the Milky Way appearing to rise like steam from the teapot's spout, then sweep upward across the sky toward the Summer Triangle near the zenith and finally arc back down through the "W" of Cassiopeia in the northeast before reaching the horizon again.

3. Use the **Hand Tool** to drag the sky in a manner that allows you to follow the path of the Milky Way as described in the previous paragraph.

You can see that the Milky Way forms a continuous band of varying width across the sky from horizon to horizon. In fact, the Milky Way completely encircles the sky as seen from Earth.

4. Select the view **Milky Way-A2** under **Go/Observing Projects/Milky Way**.

The view is the same as that of the previous sequence except that the horizon has been removed and replaced by a gray line indicating the horizon position.

5. Use the **Hand Tool** to drag the view and follow the complete path of the Milky Way around the sky.

Question 1:   Does the Milky Way encircle the sky as a continuous band or is it discontinuous, disappearing and reappearing at intervals?

6. Change the outlines of the constellations to **Astronomical Stick Figures** by opening the **View Options** pane, expanding the **Constellations** layer, clicking the **Stick Figures Options** button that appears when the cursor is positioned over **Stick Figures** and changing the **Kind** to **Astronomical** in the drop box.
7. Use the **Hand Tool** to drag the view and again follow the complete path of the Milky Way around the sky. As you do so, observe the size and brightness of the Milky Way and use your observations to answer the questions that follow.

> **TIP:** You can adjust the font and color of constellation labels in the **View Options** pane. Move the cursor over the **Labels** item under the **Constellations** layer and click the **Label Options...** button that appears.

Question 2:   Toward which constellations does the Milky Way appear broadest?

Question 3:   In the direction of which constellations does the Milky Way appear narrowest?

Speculate on your observations. The fact that the Milky Way encircles the sky as seen from Earth suggests that the Solar System is somewhere inside the Milky Way Galaxy. If this were not the case, then the Milky Way would appear discontinuous, confined to a particular direction in the sky. The fact that the Milky Way appears as a relatively narrow band suggests that the Milky Way Galaxy has a flat structure. From observations of other galaxies, we know that such a flat, disklike structure is characteristic of spiral and barred spiral galaxies. Consequently, we can deduce that the Milky Way Galaxy is a type of spiral galaxy. If it were elliptical or irregular, stars would be dispersed across the sky instead of being concentrated in such a narrow band. If this deduction is correct, then the Milky Way Galaxy must also have a central bulge. As seen from our position on Earth inside the Galaxy, this bulge would likely reveal itself as a broader region in the Milky Way.

Question 4: Using the logic of the previous paragraph and your answer to question 2, toward which constellation(s) must an observer on Earth look to see the center of the Milky Way Galaxy?

Question 5: Toward which constellation would an observer on Earth need to look in order to look directly out of the Galaxy, away from the center, along the plane of its disk?

8. Select the view **Milky Way-A2** under **Go/Observing Projects/Milky Way.**
9. Press the **K** key on the keyboard to remove the constellation outlines and labels.
10. Change the time step to 3000x and follow the position of the Milky Way along the horizon line as time advances. Observe the variable angle that the Milky Way makes to the horizon at different times.

The reason for the variable angle and rising point of the Milky Way is that the plane of the Galaxy is inclined to the equatorial plane of the Earth.

11. Select the view **Milky Way-A3** under **Go/Observing Projects/Milky Way.**

The lines in the view represent three significant planes. The red line is the Celestial Equator and represents the plane of the Earth's equator. The green line is the ecliptic, the plane of the Earth's orbit around the Sun. The blue line is the Galactic equator, the plane of the disk of the Milky Way Galaxy.

12. Use **Edit/Find** to find and center the star **Mizar** in the view.

Three points are marked in the sky, the North Celestial Pole, the North Ecliptic Pole, and the North Galactic Pole. These points represent the ends of the axes perpendicular to these flat planes in space.

13. You can estimate the angle at which the Earth's rotation axis is inclined to the plane of the Galaxy by using the **Hand Tool** to measure the angular separation between the star Polaris (near the NCP) and the North Galactic Pole (or a point very close to it).
14. You can estimate the angle at which the ecliptic is inclined relative to the Galactic plane by measuring the angular separation between the star 31 Comae Berenices near the North Galactic Pole and the North Ecliptic Pole. (If you cannot get the angular separation tool to activate on 31 Comae Berenices then you can measure from Alkaid, the star on the end of the Big Dipper's handle, first to the North Galactic Pole then to the North Ecliptic Pole, and add the two readings.)

Question 6:   What is the approximate angle between the galactic plane and the plane through Earth's equator (i.e., the celestial equator)?

Question 7:   What is the approximate angle between the plane of the Galaxy and the plane of the Solar System as represented by the ecliptic?

B. Details of the Naked-Eye Structure of the Milky Way

You probably noticed some interesting dark encroachments into the band of the Milky Way in the previous observations. The following sequences will explore examples of these structures.

15. Select the view **Great Rift** under **Go/Observing Projects/Milky Way.**

In this view to the south, again from Cornwall, Canada, at midnight on June 21, the Milky Way stretches diagonally across the sky. Beginning at the star Deneb in Cygnus and extending to the star nu Ophiuchi (ν Oph) in the lower right, the Milky Way is split by a dark intrusion known as the Great Rift. The northern part of this structure is known as the Cygnus Rift and also as the Northern Coalsack for a reason you will discover shortly.

The Great Rift is representative of a broad class of objects within the Galaxy called dark nebulae. Dark nebulae reveal themselves by obscuring the light of a radiant background such as the rich star fields of the Milky Way. As you can see, the Great Rift creates the illusion that the Milky Way splits into two branches.

Dark nebulae range in size from relatively small, almost spherical Bok globules to immense clouds of gas and dust such as the Great Rift. Modern estimates suggest that the mass of the gas and dust contained in the Great Rift is equivalent to 1 million Suns!

16. Measure the angular separation between Deneb and nu Ophiuchi (ν Oph). Use this measurement to estimate the approximate angular extent of the Great Rift.

Question 8:   What is the approximate angular extent of the Great Rift in our sky?

For Southern Hemisphere observers, an equally impressive dark nebula is the Coalsack near the Southern Cross. Like its northern namesake, the Coalsack is easily seen and appreciated with the naked eye, standing in such stark contrast to the rich star fields of the southern Milky Way that it gives the illusion of being darker than the rest of the sky.

17. Select the view **Coalsack** under **Go/Observing Projects/Milky Way.**

The Coalsack, named for its darkness and its shape, is visible below the Southern Cross asterism.

18. To identify the Southern Cross asterism, open the **View Options** pane and click the **Stick Figures Options** button that appears when you position the cursor over the **Stick Figures** item under the **Constellations** layer. In the dialog window, choose **Asterisms** in the **Kind** drop box and activate the **Labels** option. Press the **K** key to toggle the constellation and label display off again.
19. To get a better view of the Coalsack, center it in the view and then zoom in to reduce the **Field of View** to about **9°.**
20. Estimate the angular size of the Coalsack from the proportion of the field of view that it fills or choose a star near its center and use the **Hand Tool** to estimate the average radius of the dark nebula.

Question 9: Assuming a basically circular shape for the Coalsack, what is the approximate angular area of this dark nebula in the sky in square degrees?

> Giant molecular clouds are discussed in Section 20-7 of Freedman and Kaufmann, *Universe,* 7th Ed.

Dark nebulae are visible evidence for the existence in the Galaxy of interstellar matter. Roughly half of this interstellar matter is in the form of immense, gravitationally bound structures known as giant molecular clouds. These vast clouds typically contain a mass of more than 10,000 Suns, primarily in the form of hydrogen and helium gas, plus a small fraction of heavier elements and dust. The dust consists of tiny, solid particles of graphite or silicates.

The interstellar medium is extremely rarefied. Despite its low density, its pervasiveness in the vast dimensions of the Galaxy lead astronomers to estimate that the matter of the interstellar medium amounts to approximately 10% of the total mass of the Milky Way Galaxy.

C. H I and H II Regions

The dark nebulae are often called H I regions (pronounced "H one"). H I is the symbol used to describe un-ionized hydrogen, and these nebulae are cool (about 100 K) and the hydrogen they contain is therefore neutral rather than ionized.

Astronomers have learned that dark nebulae are often the birthplaces of stars. If some of these newborn stars are hot, massive, O and B stars, their ultraviolet light ionizes the hydrogen in the part of the H I region surrounding them to form an H II region (pronounced "H two," this symbol describing ionized hydrogen). Because of the energy absorbed from the ultraviolet light, H II regions are hot, typically about 10,000 K. Subsequent recombination of the electrons and ions causes H II regions to

> The Balmer series of emission lines is discussed in Section 5-8 of Freedman and Kaufmann, *Universe,* 7th Ed.

emit at visible wavelengths, predominantly in the Balmer Hα emission line in the red region of the spectrum. Because of this emission, H II regions are also called **emission nebulae.**

The H I regions surrounding H II regions often appear as silhouettes. They can also become visible by reflecting the light of hot, young nearby stars. Since the dust grains within these cold H I clouds preferentially scatter light of shorter wavelength, these **reflection nebulae** shine with a blue color.

21. Select the view **M8, M20, and M21** under **Go/Observing Projects/Milky Way.**

> This M designation refers to the Messier Catalog of deep-sky objects, originally compiled by the French comet-hunter Charles Messier in 1774 to avoid erroneous identification of these fixed object as comets. The Messier Catalog contains many of the showpiece deep sky objects visible with small telescopes.

There are two different H II regions, M8 and M20, within this 5° field of view, in addition to the open cluster of stars designated M21.

22. Zoom in on each of the three M objects in turn.

M8, the Lagoon Nebula, near the bottom of the view, is an emission nebula. M20, the Trifid Nebula gets its name from the webs of dark nebulosity that trisect the H II region. The smaller, blue nebulosity in the upper part of this nebula is a reflection nebula.

23. Select the view **M42** under **Go/Observing Projects/Milky Way**.

> For a magnificent view of M42, see Figure 20–1
> of Freedman and Kaufmann, *Universe*, 7th Ed.

M42, the Great Orion Nebula, is a large, prolific stellar nursery, a region of active star formation within a giant molecular cloud whose total mass is estimated to be approximately 500,000 times that of the Sun. About 1600 light-years from Earth, M42 lies on the edge of its parent giant molecular cloud, which in turn is but one of a vast system of such clouds found in this part of the Galaxy. Near the central portion of M42 lies the Trapezium, named for the four brightest components of the system of recently formed O and B type stars that are providing the radiation that lights up the surrounding nebulae.

24. In this magnified view of M42, identify a member star of the Trapezium and center it in the view. Then change the Field of View to about 10' to view the four stars that give the Trapezium its name.
25. Change the Field of View back to 1°. Position the cursor over the Trapezium and use the angular measurement tool to estimate the angular radius of M42.

Question 10: What is the approximate average angular radius of M42?

Question 11: At its distance of 1600 light-years from Earth, what is the approximate physical radius of M42 in light-years?

H II regions make splendid targets for small and large telescopes alike, forming large, bright canvasses against which the marvelously complex morphology of the surrounding, colder dark H I cloud is frequently silhouetted. When viewed (and particularly when photographed) through a telescope, the combination of newborn stars, dark nebulae, reflection nebulae, and H II regions make the stellar nurseries of the Galaxy provocatively beautiful.

26. Select the view **Horsehead** under **Go/Observing Projects/Milky Way**. Zoom in on the field to see how this silhouette of an H I region against a bright H II region got its name.

This region of the sky is a perfect example of the effect of dense dust clouds in obscuring more distant stars. The region of the sky to the left of the bright emission nebula apparently contains far less stars than does the region of the nebula. This is because this part of the nebula is a dense H I region containing large quantities of absorbing dust, which has obscured the background stars in this region of the sky.

D. Evolution of H II Regions

> See Figures 20–1, 20–2, and
> 20–10 in Freedman and
> Kaufmann, *Universe*, 7th Ed.

Stellar winds from newborn stars at the core of an H II region create shock waves that can trigger new regions of star formation within the surrounding dark nebula.

27. Select the view **M42** again under **Go/Observing Projects/Milky Way** and zoom in to a field of view about 40' wide.

In the magnified image of M42, you can see such a shock wave, which appears as a slightly brighter edge in the boundary between the blue reflection nebula and red H II region in the lower left part of the image.

The lifetime of a typical H I region is only several million years because its gas and dust are consumed over time in the formation of new stars and H II regions. Because the gas in an H II region is hot, it is not gravitationally bound and expands outward. Eventually, very little gas and dust remains, leaving only a cluster of newborn stars in its wake.

> 28. Select the view **M16** under **Go/Observing Projects/Milky Way**.

M16, popularly called the Eagle Nebula because of its shape, is an H II region approximately 7000 light-years from Earth toward the inner part of the Galaxy. The members of the embedded cluster of stars are estimated to be less than 1 million years old. Section 20-6 and Figure 20-16 in Freedman and Kaufmann, *Universe*, 7th Ed., discusses this nebula and shows an inset image from the Hubble Space Telescope of the detailed structure in a portion of the nebula. In the Hubble image, protostars can be seen, just emerging from their enveloping cocoons of gas and dust.

E. Open Clusters

Open clusters are the progeny of H I regions. On the whole, the member stars of an open cluster are gravitationally bound to one another. Over time however, some members of the cluster, perhaps nudged by other stars in the cluster, escape the gravitational hold of their siblings. This reduces the overall mass of the cluster as a whole, thereby decreasing its gravitational hold on the members that remain. Thus, an open cluster gradually "evaporates" over a time of several million to several billion years as its member stars disperse themselves through the Galaxy. Nevertheless, virtually all of the stars in the disk of the Galaxy were once members of such "families" of stars.

Several open clusters are visible to the naked eye and have been known since antiquity. These include the Pleiades (the Seven Sisters) and the Hyades, both in Taurus, and Praesepe (the Beehive) in Cancer.

> 29. Select the view **M45** under **Go/Observing Projects/Milky Way**.

In this telescopic view of the Pleiades, the brighter members of the cluster gleam brilliantly, their light reflected in residual strands of the gas and dust from which they were born. As with all reflection nebulae, the reflected light is predominantly blue.

> 30. Select the view **M36, M37, and M38** under **Go/Observing Projects/Milky Way**.
> 31. To label all of these clusters, open the **View Options** pane, expand the **Deep Space** layer, and click the **Labels** box next to **Messier Objects**.

Three open clusters, all in the constellation Auriga, are visible in this field of view, M36, M37, and M38. All three clusters are approximately 4000 light-years from Earth suggesting that they might all have been born within the same H I region.

> 32. Zoom in on each of the three open clusters in turn.

F. Globular Clusters

> Globular clusters are discussed in Section 21-3 and illustrated in Figures 21-10 and 25-5 of Freedman and Kaufmann, *Universe*, 7th Ed.

The Galaxy contains another type of star cluster, the globular cluster, which gets its name from its strongly condensed, distinctly globular shape. Unlike open clusters, which are concentrated in the spiral arms of the Galaxy's disk, globular clusters are distributed in a vast and roughly spherical halo around the Galaxy's central bulge.

Globular clusters are much older and much richer than their open counterparts, possessing tens of thousands to millions of member stars. The spectra of the stars in globular clusters show them to be metal-poor (to an astronomer, any element heavier than helium is a metal!), their chemical composition being not far removed from the abundance of elements produced in the Big Bang. In contrast, the stars of open clusters, and virtually all of the other stars in the galactic disk, have metal-rich spectra. Another feature of globular clusters is the virtual absence of dust and gas within the cluster, in distinct contrast to open clusters.

From their age and distribution within the Galaxy, globular clusters must have formed from the central condensation of matter in the protogalaxy that became the Milky Way. Globular clusters are spectacular sights when seen through a telescope.

33. Select **Omega Centauri** under **Go/Observing Projects/Milky Way.**

The view shows a telescopic image of the finest globular star cluster in the sky, Omega Centauri, catalogued as NGC 5139. This cluster of more than 1 million stars is approximately 16,000 light-years from Earth in the direction of the constellation Centaurus. As you can see, globular clusters are aptly named.

34. Use the angular measurement tool to measure the angular radius of Omega Centauri.
35. Open the Info pane to learn more about this fascinating object.

Question 12: What is the approximate angular radius of the globular cluster Omega Centauri?

Question 13: Given that the distance to this object is 16,000 light years, what is the radius of Omega Centauri in light years?

Question 14: If we assume that Omega Centauri is spherical and that it contains 1 million stars, what is the average density of the cluster in units of stars per cubic light-year? (*Hint:* The volume of a sphere of radius R is $4\pi R^3/3$.)

G. Planetary Nebulae and Supernova Remnants

> Section 22-3 in Freedman and Kaufmann, *Universe*, 7th Ed., describes the death of low-mass stars and their evolution into planetary nebulae.

After stars are born in the stellar nurseries of H I regions, their subsequent lifetime on the main sequence, as well as their destinies when the fuel in their cores is exhausted, depends upon their initial mass. **Low-mass** stars, those whose initial masses on the main sequence were less than eight solar masses, die by gently ejecting their outer layers into the interstellar medium to produce so-called **planetary nebulae.** The ejection is caused by bursts of nuclear energy in shells of matter near the boundary of the star's core.

> Planetary nebulae provide compelling telescopic images as seen, for example, in Figure 22–6 of Freedman and Kaufmann, *Universe*, 7th Ed.

The material of these ejected shells consists of hydrogen and helium, as well as atoms of heavier elements, such as carbon and oxygen, that were produced in various phases of the star's evolution after it left the main sequence. These atoms absorb radiation emitted from the dying parent star and subsequently re-emit it in the characteristic wavelengths for that element, producing the glowing shell of gas that we see as a planetary nebula.

The ejection of the outer part of the star leaves behind the very hot central core, which is now the central "star" of the planetary nebula. By this time, all nuclear burning within the star has ceased, and it contracts to become a **white dwarf**. The concentric shells of gas forming the planetary nebula gradually dim and, after approximately 50,000 years, become dispersed into the interstellar medium.

> 36. Select the view M57 under **Go/Observing Projects/Milky Way**.
> 37. Open the **Info** pane and expand the **Description** layer to learn more about M57.

In this view of M57, the Ring Nebula, the white dwarf star, the core of the star that produced the planetary nebula, is visible at the center of the doughnut shaped shell of ejected gas.

Another fine example of a planetary nebula is M27, the Dumbbell Nebula, which is more than 1250 light-years from Earth in the direction of the constellation Vulpecula.

> 38. Select the view M27 under **Go/Observing Projects/Milky Way**.
> 39. Open the **Info** pane and expand the **Description** page to learn more about this planetary nebula.

Again, a white dwarf star is visible at the center of the nebula.

For **high-mass** stars whose mass on the main sequence exceeds eight solar masses, death arrives in a more spectacular fashion. These stellar heavyweights end their lives in gigantic thermonuclear conflagrations. In these **supernovae**, the star's core collapses to become a neutron star or a black hole, while the rest of the star is ejected violently, scattering debris into the interstellar medium.

> Section 22–6 of Freedman and Kaufmann, *Universe*, 7th Ed., discusses supernovae.

As this debris blasts into the dust and gas of the interstellar medium at supersonic speeds, it excites the surrounding gas and causes it to glow. The visual manifestation of this collision is called a **supernova remnant**. In contrast to the gentler ejections that produce the small planetary nebulae around dying stars of lower mass, supernova explosions are so large and violent that the debris hurled into space continues to disperse at supersonic speed for tens of thousands of years. Consequently, many supernova remnants cover large areas of the sky.

> 40. Select the view **Veil Nebula** under **Go/Observing Projects/Milky Way**. Open the **Info** pane to learn more about this supernova remnant.

The Veil Nebula is approximately 2500 light-years from the Earth. It appears as two separate arcs in the sky in the direction of Cygnus.

> 41. Select the view **M1** under **Go/Observing Projects/Milky Way** and open the **Info** pane for this object.

M1, the Crab Nebula is the remnant of a supernova, the light from which arrived at Earth in 1054 A.D. We know this because Chinese astronomers of the time recorded this explosion as a star that became visible in daylight for a few days in that year.

42. Use the angular measurement tool to find the largest radial extent of this nebula (i.e., angular distance from the center of the nebula to its farthest outer border).

Question 15:   What is the largest angular radius of M1?

Question 16:   Given that the distance to M1, the Crab Nebula, is 6500 light-years, what is its largest radial extent in light years?

Question 17:   What average expansion rate does this radial extent correspond to, in light-years per year?

Question 18:   How fast is the debris from the Crab Nebula expanding into the interstellar medium in kilometers per second?

H. Distribution of Objects in the Milky Way Galaxy

It is instructive to examine the distribution in the Milky Way Galaxy of the various types of objects that were surveyed in the previous sections of this project.

43. Select the view **Milky Way-H** under **Go/Observing Projects/Milky Way**.

The view is the same as that of Section A. The southern sky is shown from Cornwall, Canada, at midnight on June 21. However, the horizon, stars, planets, moons, asteroids, and comets have all been removed from the display, leaving only the Milky Way and a line representing the galactic equator.

First, you can examine the location of open clusters in our Galaxy.

44. Use the **Hand Tool** to drag the view so that you have surveyed the entire celestial sphere.
45. Open the **View Options** pane, expand the layer labeled **Deep Space** and then the **NGC-IC Database**.
46. Click on the **NGC-IC Database** and select **Open Cluster** to display many of the open clusters in the Milky Way.
47. Use the **Hand Tool** or scrollbars to survey the sky, noting the distribution of open star clusters in the Milky Way.
48. Select **View Options/Constellations/Labels** and survey the sky once more, noting the constellation or constellations in which the density of open star clusters appears highest.

Question 19:   Relative to the galactic equator, where are most of the open clusters in the Galaxy found?

Question 20:   Toward which constellation or constellations do most of the open clusters appear?

49. Deselect the **Labels** option under the **Constellations** layer in the **View Options** pane.
50. Deselect the **Open Cluster** option under the **Deep Space/IGC-IC Database** layer of the **View Options** pane and **select** the **Globular Cluster** option.

51. Survey the distribution of this sample of globular star clusters in the Milky Way Galaxy.
52. Select the **Labels** option under the **Constellations** layer of the **Views Options** pane and repeat your survey of the sky, noting in which constellation or constellations globular clusters occur most frequently and least frequently.

Question 21: Are globular clusters found with the same distribution as open clusters?

Question 22: Toward which constellation or constellations do most of the globular clusters appear?

Question 23: Toward which constellation or constellations is the distribution of globular clusters a minimum?

These methods are discussed in the Distance Ladder project elsewhere in this book.

In 1920, Harlow Shapley published the results of his observations of RR Lyrae stars in 93 globular clusters. RR Lyrae stars demonstrate a period-luminosity relationship that allows astronomers to determine their absolute magnitude from which, with a measurement of their apparent magnitude, their distances from Earth can be determined. Knowing their positions in the sky as well as their distances, Shapley was able to analyze the three-dimensional distribution of globular clusters. He noted that the clusters were distributed in a relatively spherical halo centered over a point lying in the direction of the constellation Sagittarius. With the reasonable assumption that this distribution surrounds the central gravitational mass of the Galaxy, Shapley concluded that the center of the Milky Way Galaxy was approximately 50 kiloparsecs from the Sun in the direction of Sagittarius. Shapley turned out to have overestimated this distance by a factor of about two because he failed to account for the effect of interstellar extinction, in which intervening galactic dust artificially reduces an object's apparent brightness and consequently inflates its derived distance. Modern observations suggest that the radius of the Galaxy's disk is about 25 kiloparsecs and that the Sun is approximately 8 kiloparsecs from the center.

Question 24: What is the approximate diameter of the disk of the Milky Way in light-years?

Question 25: How far from the center of the Galaxy is the solar system in light-years?

I. The Location and Orientation of the Solar System in the Galaxy

In this section, you will use *Deep Space Explorer*™ to gain a broader perspective on the Milky Way Galaxy and the position of the Solar System within it.

53. Exit *Starry Night*™ without saving changes to the files.
54. Launch *Deep Space Explorer*™.
55. Click the **Home** button in the control panel.

The view shows a simulated image of the Milky Way Galaxy as it might appear from a distance of 110,000 light-years. Notice the grouping of bright stars centered in the image about two-thirds of the way from the center of the Galaxy in the 6 o'clock direction. This clump of stars includes the Sun and its nearest neighbors, many of which are the familiar stars that form the constellations as seen from Earth.

56. Use the Location Scroller to drag the view upward until the line of sight is along the plane of the disk of the Milky Way (i.e., the Milky Way appears edge-on). If necessary, move the image left or right to spin the image of the galaxy so that the group of stars representing the Sun's local neighborhood is in a direct line with the center of the Milky Way.

57. Use the ▼ button to zoom in on this view until the distance from the Sun as indicated in the Distance Indicator at the bottom of the view is about 0.2 light-years. Carefully examine the pattern of stars near the Sun and look for a constellation or asterism that lies in the direction of the center of the Milky Way Galaxy. (*Hint:* Look for the Fish Hook asterism to the right of the Sun and the Teapot asterism immediately below the image of the Sun.)

58. Click the **Home** button. Again, use the Location Scroller to drag the view so that the plane of the Galaxy is seen edge-on. Then spin the Galaxy so that the center of it lies directly between the current location and the clump of stars representing the Sun's local neighborhood of stars.

59. Use the ▼ button to zoom in on this view, passing through the center of the Galaxy toward the Sun until the distance from the Sun is about 0.2 light-years. Try to identify constellations in this view. (*Hint:* Look for the Orion constellation.)

Question 26: Toward which constellation or constellations do you need to look if you want to look at the outer reaches of the Milky Way that lie opposite its center, as seen from the Earth?

Question 27: From the view of the Galaxy that you obtained upon completing step 20 above, can you explain why astronomers have named the spiral arm of the Galaxy in which the solar system is found the Orion Arm?

You can use *Deep Space Explorer*™ to gain another perspective on the orientation of the plane of the Milky Way relative to the ecliptic that you explored previously.

60. Click **Home.**

61. Use the Location Scroller to adjust the view so that the disk of the Milky Way is seen edge-on and is horizontal across the view. Spin the Galaxy with the Location Scroller so that the Sun and its neighboring stars are between your location and the center of the Milky Way.

62. Zoom in to a distance of about **2 AU** from the Sun. In this view, the orbits of the planets, including that of the Earth, are visible. Use the Location Scroller again to adjust the view slightly so that the ellipse representing the Earth's orbit closes to become approximately a straight line. Use this line, plus the fact that in this view the plane of the galaxy is approximately horizontal, to estimate the angle between the ecliptic and the plane of the galaxy.

> **TIP:** To adjust the plane of the Galaxy to a horizontal position, move the Location Scroller in small circles clockwise or counterclockwise as necessary.

Question 28: What is the approximate angle between the ecliptic plane and the galactic plane?

63. Click **Home.** Take a few minutes to observe the Milky Way from various points of view. Use the Location Scroller to view the Galaxy face on and look for evidence in this simulated image of dark clouds of gas and dust interspersed among the spiral arms. Note that gas clouds outline the spiral arms and that few individual stars appear within the arms in this view. Note also the position of the Sun and local neighborhood of stars within the Galaxy.

Your observations of the previous step revealed that the disk of the Galaxy is quite flat. Modern observations indicate that the disk of the Galaxy is only approximately 0.6 kpc (2000 light-years) thick and the central bulge has a diameter of approximately 2 kpc (6500 light-years).

J. The Evolution and Spiral Structure of the Galaxy

The disk shape of the Galaxy suggests that the immense cosmic nebula from which the Milky Way formed had some overall rotational component prior to condensing under its own gravity. As the protogalaxy that evolved into the Milky Way condensed out of the foam of the Big Bang, its overall rotation speed increased in order to conserve angular momentum. The increased rotational speed caused material to flatten into a plane perpendicular to the rotation axis of the overall mass.

All of the stars and matter of the Galaxy rotate around its central gravitational mass. The Sun and its system of planets, for example, revolve about the center of the Galaxy with a speed of about 220 km/s and complete one orbit in about 220 million years.

Question 29: Current estimates suggest that the Sun is approximately 5 billion years old. In that time, how many orbits of the Galaxy has it completed?

Question 30: Using a figure of 200,000 years as the age of the human species, through what angle around the center of the Galaxy has the Sun carried us since the dawn of humanity?

Modern radio observations and Doppler studies of the Milky Way Galaxy in the 21-cm radio emission line of hydrogen show that all of the stars, dust and gas in the Galaxy move around its center in the same general direction and with roughly the same speed throughout most of the Galaxy's disk. This is different from the planets, which obey Kepler's Laws and move more slowly, the larger are their respective orbits. This finding suggests that much of the mass of the Galaxy is contained within a vast spherical halo around its center and takes the form of cold, dark matter.

As we have learned in the previous exercise and in Section A above, the most vigorous star formation occurs in the spiral arms of the galaxy's disk. A large proportion of the newborn stars emerging from these H I regions are hot, high-mass O and B Population I stars. Within the Galaxy's central bulge and also distributed in a spherical halo around its center, the proportion of cool red–main-sequence and highly evolved post–main-sequence Population II stars increases. This overall distribution of the two populations of stars is consistent with their assumed ages. The oldest stars of the Galaxy must already have been forming in the roughly spherical central concentration of mass in the gravitationally collapsing protogalaxy even as its continued contraction gradually spun it up and flattened it out. Hence, the cool red Population II stars surrounding the center of the Galaxy are among its firstborn.

64. Click **Home.**
65. Click the **Settings** tab in the Navigation Bar and, in the Settings, adjust the sliders labeled **Visibility range, Scale of close galaxies, Color saturation,** and **Brightness** all the way to the left.

This view simulates the appearance of the Galaxy, as it might look from 110,000 light-years out of the plane of its disk against a background uncluttered by distant galaxies. Notice the blue color of the spiral arms compared to the yellowish hue in the region of the central bulge. Population I, O, and B type stars are emerging from H I regions within the spiral arms of the disk, whereas the central bulge and halo contain a larger proportion of ancient, highly evolved and reddish post–main-sequence stars and cool, red, low-mass Population II stars.

66. To see an even more colorful example of this spatial distribution of stellar populations in a spiral galaxy, select **Edit/Find** from the *Deep Space Explorer*™ menu to find "**M100.**"
67. Use the Location Scroller to observe M100, a galaxy approximately 54 MLY from the Earth, from different viewpoints.
68. As you admire M100, you should be aware that a significant amount of the "blueness" of the spiral arms seen in telescopic images of other spiral galaxies in the universe, such as M100, results from the scattering of starlight within the galaxy's disk by the reflecting dust contained within the vast clouds of gas that define their spiral structure.

While the previous arguments help explain the distinct distribution of old Population II stars and young Population I stars within the Galaxy, the question of why new star formation in the present Galaxy is largely confined to the region of the spiral arms remains.

For star formation to begin in a region of a dark nebula or giant molecular cloud, some disturbance must initiate density fluctuations within the dust and gas in the cloud. A possible trigger we have suggested previously is shock waves from nearby stellar explosions. Triggers such as this appear to initiate star-forming regions in which the proportion of hot, high-mass O and B stars is much lower than in star-forming regions within the spiral arms generally. The reasons for this difference are not completely known, but one can speculate that the prodigious production of high-mass O and B type stars more typical of star forming regions in the spiral arms is the result of collision of extremely vast quantities of interstellar matter, for example, the collision of two giant molecular clouds each containing several hundred thousand solar masses of material. The shock waves and matter expelled in planetary nebulae and supernovae that disturb the homogeneity of the interstellar medium must be relatively subtle effects in comparison. Nevertheless, if these collisions of huge structures of gas and dust arose only from their individual motions within the Galaxy, we would expect the central bulge of the Galaxy to show a similar amount of star formation to the spiral arms.

The formation of spiral arms in galaxies is discussed in Section 25–5 of Freedman and Kaufmann, *Universe,* 7th Ed.

One model that astronomers have developed to explain the high rate of star formation in the spiral arms of the Galaxy uses the concept of density waves. This model proposes spiral density waves that travel through the interstellar medium of the Galaxy. These density waves travel more slowly around the Galaxy than the stars and interstellar matter. Consequently, matter and stars pile up behind the slower moving density waves. The resulting ripple of increased density gives birth to large H I regions, which in turn produce stars.

As the clusters of newborn stars continue to revolve around the center of the Galaxy with the same speed as the rest of the stars and matter of the Galaxy's disk, they gradually outpace the slower moving density wave and are no longer associated with it. The high-mass O and B stars born within these clusters, however, have relatively short lifetimes compared to their lower-mass siblings further down the spectral sequence, and have exploded and disappeared by the time the cluster leaves the density wave. Consequently, it is the distribution of the hot O and B type stars that most closely reflects the position and structure of the underlying density waves.

THE LOCAL GROUP OF GALAXIES

22

Many centuries ago, Greek astronomers debated the question of the position of the Earth within the observable universe. They concluded that the Earth was at the center of this universe. Our increasing base of knowledge has slowly downgraded this status, from the central object of the universe, to an average planet in orbit around a rather ordinary star, accompanying many other stars in a galaxy of stars among billions of such galaxies.

As part of this progression, significant advances in the sixteenth and seventeenth centuries led to a rationalization of the solar system in which the Sun replaced the Earth as the dominant object, but this Sun was still considered to occupy a central position in the overall universe. Further observations revealed the true nature of the Milky Way as a galaxy of stars, and the study of clusters of stars within this galaxy led to the realization that the Sun was not even at the center of this galaxy. Furthermore, observational evidence in the early part of the twentieth century began to show that the Milky Way was not the only large structure of its kind and that the universe was populated by many billions of such structures distributed over vast distances. More recent work has shown that these distant galaxies are assembled into clusters and superclusters of galaxies that form vast walls surrounding relatively empty regions or voids in space. These structures are explored in more detail in "The Large-Scale Structure of the Universe" project later in this book.

The M designation comes from a catalogue of about 110 diffuse objects assembled by Charles Messier in the late 1700's to help comet hunters like himself to avoid erroneous identification of these objects as comets.

The **Milky Way** is now known to be a part of a small and mixed collection of galaxies in a loose cluster known as the **Local Group**. This project will use *Deep Space Explorer*™ to investigate the size, makeup and distribution of this group of galaxies.

The Local Group consists of more than 40 galaxies, at least one of which, M31, the Andromeda Galaxy, is sufficiently bright to be visible with the naked eye from a dark site upon Earth. The Local Group contains several large galaxies, such as the Milky Way and the Andromeda Galaxy, but there are many smaller and fainter companions, some of which have been discovered only very recently.

In the sections that follow, you will use *Deep Space Explorer*™ to look at and even "visit" many members of the Local Group, something that is impossible to contemplate in the real world. Even light, traveling at the fastest possible speed in our universe, takes hundreds of thousands and even millions of years to cover the distances from these galaxies. *Deep Space Explorer*™ uses the best available data on a large number of galaxies to assemble a true three-dimensional field through which you can "travel," stopping to admire the view and to identify the different components of our universe as you do so.

A. Galaxy Classification

> The Hubble classification is discussed in Chapter 26, "Galaxies" in Freedman and Kaufmann, *Universe*, 7th Ed.

In this project, you will observe several galaxies and classify them according to an accepted scheme devised by Edwin Hubble, for whom the highly successful Hubble Space Telescope is named. Hubble separated galaxies into four main groups according to their appearance.

It is worth noting that *Deep Space Explorer*™ uses a revised and somewhat more complex Hubble classification scheme for galaxies in the On-Screen Information. In this project, we will adhere to the more traditional classification scheme discussed in Freedman and Kaufmann, *Universe*, 7th Ed.

The four main groups in the Hubble classification of galaxies are: S for Spirals, SB for Barred Spirals, E for Ellipticals, and Irr for Irregular galaxies. These galaxies differ not only in appearance but also in composition. For example, Spirals and Barred Spirals contain lots of dust and gas and many of their stars appear to be relatively young. Elliptical galaxies contain almost no dust and gas and the component stars appear to belong to an older population. Irregular galaxies contain mostly young stars as well as dust and gas.

A relatively flat disk of stars encircling a central bulge is the distinctive feature of **Spiral Galaxies**. The name comes from the apparent spiral distribution of stars and other matter within the disk, beyond the central bulge of these galaxies. **Barred spirals** are distinguished from other spiral galaxies by a straight bar across the central bulge, the spiral arms in the disk starting at the ends of this bar and not at the edge of the central bulge. Spiral and barred spiral galaxy groups are subclassified into three main categories, Sa, Sb, or Sc and SBa, SBb, or SBc, respectively. These three subclassifications are distinguished according to the following criteria:

   a) Sa and SBa galaxies have fat central bulges and smooth, broad spiral arms.

   b) Sb and SBb galaxies have moderate central bulges and moderately well-defined spiral arms.

   c) Sc and SBc galaxies have small central bulges and narrow, well-defined spiral arms.

Differences between Sa, Sb, and Sc galaxies may be related to the amount of dust and gas within their spiral arms. This dust and gas component is important in the production of new stars within galaxies. The approximate percentage of the mass of these galaxies in the form of dust and gas is 4% for Sa galaxies, 8% for Sb galaxies, and 25% for Sc galaxies.

Elliptical Galaxies have no disk or spiral arms. Although they are all elliptical in shape, the ellipse can vary from essentially circular to very flat. Hubble classified them in terms of their apparent flattening, from E0 to E7, in which E0 galaxies are the most circular in appearance and E7 galaxies show the most flattening.

The appearance of an elliptical galaxy in this classification scheme does not necessarily describe the true three-dimensional nature of the galaxy. An E0 galaxy might be spherically symmetric or it might be a flattened disk seen face-on from our view. A cigar-shaped galaxy seen end-on might also appear to be spherical and be classified as an E0 galaxy. This is different from the classification of spiral galaxies, where the appearance of the spiral arms indicates the viewing direction on the galaxy from our point of view. We see examples of all directions of alignment of spiral galaxies to our line of sight, so we can be fairly certain that we know the true three-dimensional shapes of these galaxies.

Elliptical galaxies contain little or no gas and dust and thus appear much less dramatic than spiral galaxies. In the spiral galaxies, blue and ultraviolet light from young, hot stars illuminates the dust and gas clouds. These stars have been formed fairly recently in the galaxy's history and outline the spiral arms, making them very distinct. In contrast, the lack of dust and gas in elliptical galaxies shows that star formation ceased in these galaxies long ago. Elliptical galaxies are composed of old, red stars and show no evidence of spiral structure. Nevertheless, elliptical galaxies come in the widest range of sizes and masses, from the largest known galaxies to very small aggregates of old stars. Often the gravity from the mass of one or two giant elliptical galaxies will dominate the motions of entire groups or clusters of galaxies.

> Hubble's tuning fork diagram is illustrated in Figure 26–9 of Freedman and Kaufmann, *Universe*, 7th Ed.

Hubble summarized his classification scheme by drawing the types of galaxies on a tuning-fork diagram in which the handle contains the elliptical galaxies, from E0 at the end to E7 near to the forks. The spiral and barred spiral galaxies occupy the two tines of the fork, arranged from Sa and SBa near the handle to Sc and SBc at the ends of the tines.

At the junction of this fork, Hubble placed another group of galaxies he called **Lenticular** galaxies. These resemble elliptical galaxies in lacking gas and dust and in showing no spiral structure but appear to have a definite central bulge. Hubble denoted these as S0 for normal lenticular galaxies and SB0 galaxies if they showed a bar.

The **Irregular** galaxies did not fit into this tidy scheme but Hubble defined two types:

Pictorial examples of the different types of galaxies can be found in Chapter 26 of Freedman and Kaufmann, *Universe*, 7th Ed. The textbook also gives a summary of the average physical properties of the three main classes of galaxies in Table 26-1.

1) Irr I galaxies, which look like undeveloped spiral galaxies and contain many young stars and lots of dust and gas.

2) Irr II galaxies, which are distorted and show no particular symmetry. They appear to have been formed by collisions with other galaxies or disturbed by violent activity within their interiors.

One further characteristic of our universe is the distribution of galaxies. These massive collections of stars and other matter also seem to congregate together in **clusters**. The number of galaxies in a cluster can vary widely. Those with few members are known as **poor clusters** and are often termed groups, while those with many hundreds and thousands of galaxies are known as **rich clusters**. Furthermore, their distribution around some center of concentration leads to a second classification. Those clusters with a distinct spherical appearance are called regular clusters, whereas those that appear to be scattered randomly over a large region of space are called irregular clusters.

The properties of these massive conglomerates of galaxies are not yet fully understood. Various anomalies have arisen in recent studies of galaxies and clusters of galaxies that have led to the postulation of dark matter to explain gravitational effects that appear to require far more mass than is provided by the obvious visible matter.

We find that our Milky Way Galaxy is a member of a poor cluster known as the Local Group. This group contains at least 40 other galaxies and extends out to a distance of approximately 10 MLY from the Milky Way Galaxy. Faint new members of this Local Group are still being discovered as astronomers develop improved observing techniques. Many members may never be discovered because our own Milky Way obscures certain areas of the sky from our view in a region known as the Zone of Avoidance. This region is explored briefly below.

With the above summary of galaxy types as a reference, you can now explore the Local Group, a poor cluster containing a wide variety of galaxies. You can use *Deep Space Explorer*™ to "visit" several of these component galaxies and classify them under the Hubble scheme. You can determine several of the physical characteristics of these member galaxies, either from the images presented in *Deep Space Explorer*™ or from the Information Table presented for each chosen galaxy.

In our exploration of this small group of galaxies, you can start from the *Deep Space Explorer*™ default reference point known as **Home**. This position is 110,000 LY above the Sun's position, which is in a spiral arm of the Milky Way Galaxy. You can peruse the view from this position before looking further afield to our near neighbor galaxies. You can then move to the more distant members of this select group of galaxies and examine their mutual spatial relationships.

The word "neighbor" invokes thoughts of some relatively close location but it is well to keep in mind that light from most of our neighbor galaxies has taken at least 80,000 years to reach us, and

Examples of images of these very distant galaxies are shown in Figure 26-31 of Freedman and Kaufmann, *Universe*, 7th Ed.

as long as 10 million years from the farthest Local Group galaxy. Thus, we see these objects as they were a long time in the past, when this light left them. They are nevertheless close compared to the distant field of galaxies in the present dataset, which extends as far as 350 MLY from us. The Local Group galaxies are certainly in our backyard compared to the most distant galaxies recently imaged by the Hubble Space Telescope and several large new telescopes in Hawaii and South America. These modern instruments have produced images from the light that left these distant galaxies up to 12 billion years ago and has only recently reached the Earth!

B. The Milky Way Galaxy

You can now begin the exploration of our region of space, beginning with an inward journey into the spiral arms of our Milky Way Galaxy and the region of the Sun's neighborhood.

1. Launch *Deep Space Explorer*™.

The default view of the representation of our galaxy shows a magnificent spiral structure, with arms outlined by gas and dust clouds illuminated by bright stars, along with many individual stars dotted along these arms. These distinct arms connect to a rather small central bulge. A bright group of stars at the center of the view indicates the Sun's location within a rather indistinct spiral arm.

> **Question 1:** On the basis of the appearance of the Milky Way Galaxy from above its spiral arm plane, what would be the Hubble classification of this galaxy?

You can use the Location Scroller to rotate this image in order to examine the structure of our galaxy more closely. This rotation will be centered upon the Sun. Thus, your movement will be as if you are moving along the surface of a sphere with a radius equal to the distance displayed at the bottom of the main view window, which is 110,000 LY for the reference position.

2. Use the Location Scroller to rotate the plane of the galaxy to an edge-on position.

TIP: It is advisable to move the Location Scroller along vertical and horizontal lines centered upon the main view window in order to move in a reproducible way.

You can see that many bright points surround the Milky Way Galaxy. Almost all of these are individual galaxies at large distances from your position. If you move the cursor over any of these points, an information box, called the Heads-up Display (HUD), will appear. The HUD displays the galaxy's designation in one of a number of catalogues. NGC refers to the New General Catalogue, IC refers to the Index Catalogue and PGC refers to the Principal Galaxy Catalogue. The HUD also gives the galaxy's absolute magnitude and its distance from your present position, which will change as you approach or retreat from this object.

These catalogues are discussed in the *Deep Space Explorer*™ User's Guide.

From the view, it is also obvious that there appears to be a region surrounding the Milky Way where there are no galaxies. This is the Zone of Avoidance mentioned above, where material in the Milky Way plane has obscured these regions of the sky. Thus, there might be, and almost certainly are, galaxies in these directions but we cannot see them from Earth. You can roll the Milky Way Galaxy like a wheel to see this region completely surrounding the Galaxy.

3. Use the Location Scroller at one end of the edge-on galaxy and move it along the plane of the Galaxy's disk to rotate the galaxy around its axis like a wheel. You may have to repeat this procedure several times to rotate the galaxy fully around this axis.

TIP: You can rotate the view around its center by moving the Location Scroller in a small circle.

As you rotate the Milky Way, you will see that the zone of avoidance extends all round the Galaxy, as expected. Because you are rotating around the Sun's position, the galactic center can be moved between your position and that of the Sun. When the bright cluster of stars representing the Sun's neighborhood is diametrically opposite to your position on the other side of the center of the Galaxy, the display of this galaxy will change to a myriad of stars.

Before exploring the neighboring galaxies of the local group further, you can take a brief journey inward to the Sun's position to get a visual feel for its position within the Milky Way.

4. Click **Home** in the control panel to return to the reference position in space.

5. Zoom in toward the solar neighborhood in the spiral arm of the galaxy. Each click will reduce your distance from the Sun by about 10% and this distance will be displayed in the Distance Indicator (DI) at the bottom of the view.

6. Continue to zoom in on the Sun's position until the distance to the Sun is about 5000 LY, when a message at the top of the window will inform you that you have entered the region of local stars. This region can be loosely defined as that region of space containing stars whose distances have been measured by the Hipparcos spacecraft.

7. Continue to zoom closer to the Sun until you reach about 250 LY.

Click the ▼ elevation button to "zoom in" and the ▲ elevation

At this point, the Sun is still too faint to be seen but several of its bright neighbors are visible, particularly Arcturus just above the center of the view and Procyon and Sirius just below and to the right of center. Canopus is almost out of the view on the far right of the window.

8. Zoom in closer to the Sun until, at about 100 LY, it appears at the center of the view, suitably labeled.

9. As you continue to zoom in step-by-step toward the Sun, you will note that stars nearest to you will move more rapidly out of view. At a distance of about 0.16 LY, the units of distance will change to Astronomical Units (AU). At about 3000 AU, the orbits of the planets are displayed and at about 2000 AU, a message appears at the top of the window to inform you that you have entered the Solar System. You can now zoom in to about 1 AU, where you will be at the Earth's distance from the Sun.

10. Use the Location Scroller to move the view around and look for familiar constellations and asterisms. Stop when you find the Big Dipper.

11. Zoom out with the Up elevation button and watch the Big Dipper closely as your distance increases. Stop zooming out when you see a noticeable change in the Big Dipper's shape.

TIP: When the cursor is positioned over a star, the HUD identifies it and gives its apparent magnitude as seen from Earth as well as the distance to the star from your current location. Clicking the right mouse button while the HUD is visible will open a contextual menu. The **Show Info** option will display a comprehensive list of physical data pertaining to the chosen star. To hide this information again, open the contextual menu and select the **Hide Info** option.

The change of shape of the Big Dipper as you zoomed out of the Galaxy in the previous step occurs because the different stars that make up this asterism are at different distances from the Sun and will show different movements because of parallax effect as your observing position changes.

Question 2: Roughly how far do you have to be from the Sun before the Big Dipper begins to lose its well-known shape?

12. Continue to zoom out rapidly from the Sun to about the same distance as the reference distance, about 110,000 LY, to experience the feeling of flying rapidly away from the Sun and out through the Milky Way into the nearby space.

C. Our Near Neighbor Galaxies

Having explored the Milky Way briefly, we will now explore the space around our home Galaxy and meet some of our neighbors.

13. Click **Home.**
14. Use the Location Scroller to adjust the view so that the disk of the Milky Way Galaxy is seen edge-on.

You will notice two somewhat diffuse and nebulous regions drifting though three-dimensional space in the lower right of the display. This "drifting motion" against the distant background of galaxies represented by the field of bright points is an effect of parallax and indicates that these are local, relatively nearby objects.

15. Identify these two objects with the HUD.

You will see that these objects are the Magellanic Clouds, prominent naked-eye objects in the skies of Earth's Southern Hemisphere. The Small Magellanic Cloud appears blue. This is the result of light from hot, young stars within this galaxy being scattered and reflected by the dust within it. The Large Magellanic Cloud shows a redder color.

In the next few steps, you can zoom in to take a closer look at each galaxy before classifying them.

16. Open the contextual menu over the Small Magellanic Cloud (SMC) and select **Center Small Magellanic Cloud.**
17. Using single steps, zoom in toward the SMC until the DI indicates that you are about 15,000 LY away. The SMC almost fills the view from this distance.
18. Use the Location Scroller to observe the SMC from various points of view, including edge-on and face-on.

Question 3:   What is the Hubble classification of the Small Magellanic Cloud?

Question 4:   Is there any evidence of spiral structure or central bulge in the Small Magellanic cloud?

19. With the view showing the SMC face on, zoom back out from the SMC until the Large Magellanic Cloud (LMC) comes into view, when your distance from the SMC is about 0.125 MLY.
20. Center the view onto the LMC.
21. Zoom toward the LMC until you are about 40,000 LY away and use the Location Scroller to view the LMC edge-on to examine its structure. Then use the Location Scroller to move the sky until the Milky Way joins the LMC and SMC in the view.

This combination of the three near neighbors of the local group makes a superb montage of objects and illustrates the three-dimensional relationship between them.

Question 5:   What is the Hubble classification of the Large Magellanic Cloud?

Question 6:   Does the LMC have any evidence of spiral arms or central bulge to suggest an association with spiral galaxies?

22. Click **Home**.
23. Use the Location Scroller to adjust the view so that the disk of the Milky Way is edge-on and horizontal.

You will have noticed that several points of light moved relative to the background as you adjusted the view in the last step. Some are outlying stars of our galaxy but others are small galaxies accompanying our own. We will look briefly at these galaxies in a moment. However, your eye might be caught by a small image that appears just below the Milky Way plane. Let us examine this fascinating object in more detail.

24. Position the cursor over the small, diffuse object just below the axis of the Milky Way Galaxy. The HUD identifies the name of this galaxy, Sagittarius, and lists its absolute magnitude and its distance from your current position in space.
25. Center the view on Sagittarius.
26. Zoom in toward Sagittarius until you are about 8000 LY away.
27. Examine the three-dimensional structure of this galaxy using the Location Scroller.

As you see, this galaxy is approximately symmetrical in all axes and appears to have no dust or gas within it. These observations should allow you to place this galaxy within the Hubble classification scheme. This type of galaxy is certainly VERY different in character from the Milky Way Galaxy and the Magellanic Clouds.

28. Open the contextual menu over this galaxy and select **Show Info**. For Sagittarius, you will note that its diameter and thickness are comparable, as is typical for this type of galaxy.

Question 7:   What is the Hubble classification of the Sagittarius Galaxy?

D. The Andromeda Subgroup of Galaxies

Before starting this next exploration, it is possible to reduce the confusion in the view by lowering the brightness of background galaxies. In this way, you will be able to see the Milky Way and its smaller satellite galaxies in relation to the other subgroup of galaxies in the Local Group, centered approximately on the Andromeda Galaxy, against a relatively blank background.

29. Click **Home**.
30. Open the **Settings** pane in the Navigation Bar. In the **Number of Galaxies** box, adjust the **Visibility Range** slide-bar to the left to reduce the number of visible background galaxies in the window. Except for a few nearby stars, this will leave only the nearest galaxies of the local group and a few bright but more distant galaxies in the view.
31. Center the view on the Milky Way.
32. Click several times on the Up button to zoom out to a distance of about 0.25 MLY from the center of the Milky Way.
33. With the cursor placed halfway up on the far right side of the view, activate the Location Scroller and drag the view toward the left. Another group of galaxies will enter the view from the lower left.

This group of galaxies is associated with the large spiral galaxy, M31, the Andromeda Galaxy, clearly seen in this view. M33, the Triangulum Galaxy, can also be seen, below and to the right of M31.

34. Select **File/Save As...** from the menu and save this configuration in a file named **Local Group** in the file folder *Deep Space Explorer*™ in the Program Files on your computer's hard drive.
35. To help you to identify the galaxies mentioned in the previous paragraph, click the **Labels** tab in the Navigation Bar. Click the check box for **Galaxy Labels**. When the box is checked, *Deep Space Explorer*™ will label some of the galaxies in the view.
36. Turn the galaxy labels off again.

There are two features of *Deep Space Explorer*™ that, when applied with the proper strategy, allow you to measure the distances between any two galaxies in the *Deep Space Explorer*™ database with ease. These features are the distance indicator (DI) and the **selection HUD (SHUD)**. As you know, the DI at the bottom of the view always shows the distance of your current viewing location in space from the Sun. Whenever the view is locked on to any object other than the Sun, the DI also displays the distance to that object from the current viewing location. The SHUD is a permanent version of the usual HUD. It displays the same information as the regular HUD but remains visible and attached to the object until it is specifically turned off. You can display only one SHUD at a time.

37. With the view locked on the Milky Way, use the elevation buttons to zoom in and out. Notice that as your location zooms toward or away from the Milky Way, the distance to the Milky Way displayed in the lower half of the DI is automatically updated.
38. Now use the Location Scroller to move the view in various directions. Notice that the distance to the Milky Way displayed in the DI does not change. (The distance to the Sun DOES change, however, because the Sun is offset from the fixed point of this view, the center of the Milky Way.)
39. Zoom in or out so that the distance to the Milky Way, displayed in the DI, is about 128,000 LY (0.128 MLY).
40. Use the Location Scroller to put the Large Magellanic Cloud in the view. Position the cursor over it and when the cursor becomes an arrow shape and the HUD for the LMC appears, click the left mouse button. This will activate the SHUD feature. When an object is selected by left-clicking over its image in the main view (or by using the Edit/Find command in the menu), the SHUD will remain attached to the object until it is deactivated with the Edit/Select None command in the menu, or until another object in the view is selected, in which case, the SHUD attaches itself to the new object.
41. With the SHUD attached to the Large Magellanic Cloud, and the view centered on the Milky Way, use the Location Scroller to alter the view and watch the distance value in the SHUD change. You will notice that, because the LMC is not the object centered in the view, its distance from your location does change as you change the view and this is reflected in SHUD.

TIP: If you have trouble specifying the object with the mouse cursor because it lies against a background rich with distant galaxies in the current view, use the Location Scroller to try to find a viewpoint in which the object is in or near the zone of avoidance. It should then be easy to specify the object with the mouse cursor so that the HUD can identify it.

The strategy that will allow you to use the DI and SHUD effectively to measure the distance between any two galaxies in *Deep Space Explorer*™ is quite simple. Suppose you have the view centered on galaxy X and you want to know the distance that separates it from another galaxy, galaxy Y. If you use the Location Scroller to adjust the view so that both galaxies are in the same line of sight, then the three points in space—your position, the position of galaxy X, and that of galaxy Y—all lie on the same line. The distance between the two galaxies is then simply the difference between the distances of these galaxies from your position. Because the distance between two galaxies must always be a positive quantity, the separation between galaxy X and galaxy Y is really the absolute value of the difference in their distances from your position. In practical terms, this means that it does not matter whether galaxy Y is in front of or behind the stationary galaxy X. If you attach the SHUD to galaxy Y, the two distances required for calculating the separation of the two galaxies are immediately available, one displayed in the SHUD, the other in the DI. You can use this technique to measure the distance between the Milky Way and the galaxies listed in Data Table 1 below.

42. Select **File/Open** from the menu and open the file you saved previously named **Local Group**. The view should be centered on the Milky Way and the DI should indicate a distance of about 0.25 MLY from the center of the Milky Way.
43. Identify M31, the Andromeda Galaxy, in the lower left of the view.
44. Activate the SHUD over M31.
45. Use the Location Scroller to adjust the view so that M31 is aligned with the center of the Milky Way. Record the distances displayed in the SHUD and the DI in Data Table 1.
46. Calculate the distance that separates M31 from the Milky Way.
47. Repeat steps 43 to 46 for each of the galaxies listed in Data Table 1. (*Hint:* It might be useful to activate the Labels tab briefly in the Navigation Bar and click the check box for **Galaxy Labels** to the identify the galaxies in the list.)

DATA TABLE 1

Galaxy	Distance to Galaxy (MLY)	Distance to Milky Way (MLY)	Difference (MLY)
M31 (Andromeda)			
M33 (Triangulum)			
Barnard's Galaxy			
SMC			
LMC			

Question 8: What are the separations between the Milky Way and the following Galaxies: M31, M33, Barnard's Galaxy, the SMC, and the LMC?

E. The Andromeda Galaxy and its Satellite Galaxies

You can now plan an expedition to visit the group of galaxies surrounding M31, several MLY away from the Milky Way, to classify and compare them with our own Galaxy.

48. Select **Edit/Select None** to deactivate the SHUD.
49. Find M31, the Andromeda Galaxy, in the view and lock on to it.
50. Zoom in step-by-step to this galaxy until the DI indicates that you are about 0.5 MLY from it.
51. Rotate your viewpoint around M31 to examine its structure. Note that M31 has a reasonable-sized central bulge and fairly distinct spiral arms.
52. Rotate M31 to an edge-on position in order to find its companion galaxies.
53. You can now examine several of M31's companion galaxies by locking on to them and zooming in toward them. You can then use the Location Scroller to look at them from various points of view to help in the classification of these galaxies. To move to another galaxy, simply zoom out again until the other galaxies come into view.

Try to classify the following short list of these companion galaxies, to be found near to the Andromeda Galaxy.

Galaxy Designation/Name	Hubble Classification
M31 (Andromeda)	
M33 (Triangulum)	
M110	
M32	
NGC 147	

Question 9: What are the Hubble classifications for M31, M33, M110, M32, and NGC 147?

F. The Location of the Local Group within the Universe

After this brief look at a few of the Local Group of galaxies, you can conclude this exploration by using *Deep Space Explorer*™ to show you where this cluster of galaxies fits into the larger scale of things. As you venture further into space, you find that galaxies tend to be found in rather thin walls. These walls surround vast voids in which few if any galaxies are found. Clusters of galaxies are found within these walls and the whole structure of deep space seems to resemble that of a collection of soap bubbles. The walls of galaxies and the more concentrated clusters and superclusters are found at the interstices of these bubbles.

You can explore this situation briefly by zooming out from the Milky Way, first noting again its relationship with the members of the Local Group, particularly the Andromeda Galaxy, before looking at its position within a nearby wall of galaxies.

54. Click **Home**.
55. Adjust the view so that the Milky Way is edge-on against the dark Zone of Avoidance.
56. Zoom out beyond the Milky Way to a distance of about 3.2 MLY. At this distance, the M31 sub-cluster will appear in the lower-left of the view.
57. Continue to zoom outward, step-by-step, noting that several galaxies in addition to the M31 cluster will move rapidly toward the center of the window, a sure sign that they are relatively close to the Local Group. At about 4 MLY, you will leave the confines of the Local Group.
58. Zoom out to about 25 MLY from the Milky Way, at which point several additional galaxies will have appeared between your position and the home galaxy.
59. You can now use the Location Scroller to rotate your viewpoint around the Sun's position in the Milky Way.

You can reduce the confusion produced by all of the distant galaxies and look at only those of the Local Group by moving the **Number of Galaxies** slider bar to the far left in the Settings pane. Adjustment of the **Brightness** slider bar can then be used to increase the brightness of the Local Group galaxies.

From this elevated viewpoint, you will see that there are many galaxies between you and the Milky Way that make up a very thin wall of galaxies. Judicious use of the Location Scroller will allow you to place yourself inside this thin structure and look edge-on through it. Moving the Location Scroller in a small circle anywhere in the window allows you to rotate the plane of this wall. It is easy to see the empty space on either side of this wall, the beginnings of two local voids.

60. Zoom out to about 75 MLY from the Milky Way. At this point, rotation of the field reveals vast superclusters of galaxies at the corners of voids that are surrounded by huge walls of galaxies.
61. Open the **Highlights** pane. In the drop box near the top of the pane, select **Filament Family**. Find and click on the check box beside the item named **GA Coma-Sculptor Cloud** in the **Supergalactic Eq Plane** layer to highlight this vast wall in which we live and move your viewpoint to see the extent of this structure.

From this viewpoint and with this association of galaxies highlighted, you can see the narrowness of this wall. Furthermore, by manipulating your point of view, you can see that this vast structure connects to one of the truly gigantic superclusters of our universe, the Virgo A cluster, at the end of our wall, at the corners of several vast voids.

G. Conclusion

You have explored the relationships between our Galaxy and its near neighbors in the Local Group of galaxies. You have been able to measure the scale of this space by estimating the distances separating some of the components of this cluster. Finally, you have been able to place yourself within the large-scale structure of the universe in a way that two-dimensional pictures could never do, to allow you to glimpse the bubblelike features that lace our local extragalactic neighborhood.

THE LARGE-SCALE STRUCTURE OF THE UNIVERSE 23

Our own Milky Way Galaxy is a gravitationally bound structure consisting of a huge flattened disk made up of about 200 billion stars and large quantities of dust and gas in spiral structures surrounding a central bulge. Our increasingly precise and penetrating observations of the universe over the past century have revealed that the Milky Way is but one of many billions of such collections of matter, many of them part of immense clusters and superclusters of galaxies. The past few decades in particular have led us to the realization that even galaxies outside these huge conglomerations are not randomly distributed in space but occur within huge walls, surrounding vast voids that contain few if any galaxies. The present view of the large-scale structure of the universe is one that resembles a collection of soap bubbles. Clusters and superclusters of galaxies appear to occupy the lines where the walls meet, the largest of them being at the corners of the voids. Furthermore, at least one of these superclusters, dubbed the Great Attractor, appears to exert sufficient gravitational influence upon galaxies, including our own Milky Way Galaxy, that these galaxies are moving collectively toward this region of space.

> Section 26–6 in Freedman and Kaufmann, *Universe*, 7th Ed., discusses clusters, superclusters and the "foamy" structure of the universe.

In this project, you will use *Deep Space Explorer*™ to venture among these vast structures, stopping at important way stations to examine the detailed and interrelated collections of galaxies that make up our corner of the universe.

A. The Milky Way and the Local Wall

You can start your journey from the Milky Way Galaxy and move slowly outward to visualize its position within the Local Group, a poor cluster of more than 40 galaxies of all types distributed over a volume inside a radius of 5 to 10 million light-years.

1. Launch *Deep Space Explorer*™.

The view shows the Milky Way from a location 110,000 light-years from the Sun. Before speeding off into deep space, it is instructive to examine the universe from this "local" position. In particular, significant regions on the sky are blocked from our view because matter within the Milky Way absorbs the light of distant galaxies, and it is helpful to look at the effect that this **Zone of Avoidance** has on our view of the sky.

2. Use the Location Scroller to move your view of the Milky Way so that it appears edge-on.

Almost all of the bright points in the view window are distant galaxies but you will note the apparent absence of such galaxies around the plane of the Milky Way. This lack of galaxies is not real, but shows our ignorance of this region of the sky because of absorption of light by material in our own galactic environment. You should keep this limitation in mind as you explore further into space.

3. Use the Location Scroller to roll the Milky Way around its axis like a wheel to demonstrate that the Zone of Avoidance extends completely around our position, as expected. Also, notice that this Zone of Avoidance is larger in extent when the view is toward the center of the Galaxy, as opposed to the view out of the Galaxy.

As you move your viewpoint, you will notice that local galaxies, such as the Large and Small Magellanic Clouds and the giant Andromeda Galaxy, move relatively rapidly across the view. This is because they are in the Local Group of galaxies and are close to the Milky Way. You can take a brief look at this group of our neighboring galaxies.

4. Click **Home** and use the Location Scroller to place the Milky Way edge-on again.
5. To reduce the confusion from distant galaxies, click the **Settings** tab and reduce the number of visible galaxies by moving the **Number of Galaxies** slide-bar to the far left in the Settings pane.
6. Click on the Up (▲) button on the control panel to zoom away from the Milky Way Galaxy until the Distance Indicator at the bottom of the view displays a distance from the Sun of about **3 MLY**. You will see the Milky Way and its near neighbor galaxies recede into the distance as the other sub-cluster of galaxies around the giant Andromeda Galaxy come into view at about this distance, in the bottom-left corner of the view.
7. Further recession to about **7.5 MLY** from the Milky Way will show a few of the outlying galaxies of our Local Group.
8. Adjust the **Brightness** of galaxies with the slider bar in the Settings pane to brighten more of these nearby galaxies. Be careful not to move this brightness adjustment too far because the many distant galaxies will confuse the view of the relatively nearby galaxies.
9. Zoom out to about **30 MLY**, where you will see many more galaxies move into your view window.
10. Use the Location Scroller in a top-left to bottom-right motion to rotate around the position of the Milky Way in the center of the view window to see how the Local Group of galaxies fits into a narrow wall of galaxies. You can use the Location Scroller to move this wall to an edge-on position and then roll it along like a wheel to explore its relationship with other structures in this region of space within 30 MLY of our position.

This wall inside which the Milky Way resides separates two voids, as shown by the lack of local galaxies in these regions. It is instructive to return to Home and repeat this journey quickly with the background galaxies visible to see how much more difficult it is to discern this wall against the rich and confusing background.

11. Click **Home.**
12. Use the Location Scroller to place the Milky Way edge-on and zoom out again to about **30 MLY** from the Sun.
13. Use the Location Scroller to rotate the wall in space, around the Sun's position.

Deep Space Explorer™ can make the identification of this feature (and many others) easy to see.

14. Click on the **Highlights** tab. Make sure that the **Show filaments/group labels** box is checked at the top of this pane and that "Filament family" is selected in the box above the object list. In the **Supergalactic Eq. Plane** list, click on **GA-Coma-Sculptor Cloud** to highlight this wall of galaxies, which extends between the Great Attractor through Coma and Sculptor, hence its name, and contains the Local Group of galaxies.
15. Use the Location Scroller to move your point of view around to view this extended wall from other positions.
16. Zoom out to about 80 MLY from the Sun and again use the Location Scroller to move the sky around to view the fine filigree structure in three dimensions.

Moving around the Sun's position at this distance, you get the sensation of moving through bubbles of space with thin walls of galaxies separating large voids. You can move the sky around to show that the wall that you have highlighted is connected to several superclusters of galaxies at the interstices of these bubbles.

17. Click on the **GA-Virgo Cluster** in the **Supergalactic Eq Plane** layer in the **Highlights** pane to outline one of these enormous clusters. Again, you can rotate the plane of the sky to see the relationship between the wall and this cluster, in the corner of a void.
18. Click on the **Supergalactic Eq Plane** to highlight the whole of a giant wall. (This highlighting illuminates the whole list of smaller structures in the section of the table. This wall is used to define an "Equatorial Plane" for reference when discussing this region of extragalactic space.)
19. Click off the "Show filaments/group labels" to reduce the confusion from these individual labels.
20. Rotate the sky again to demonstrate that this is a relatively flat wall by finding the direction from which this structure is edge-on to your viewpoint.
21. Finally, zoom out to about **200 MLY** from the Sun. You can rotate the sky again and roll this wall like a wheel to see that the wall has linear structures within it and that narrow subsidiary walls are attached to this vast wall and spread away from it in all directions like the branches of a Christmas tree.

It is perhaps somewhat overwhelming to remember that each bright point in this image is a galaxy in its own right, each one containing many thousands to millions of stars, and that you are looking (in simulation, admittedly) at the universe on a truly vast scale.

B. Superclusters of Galaxies

As you saw at the end of the last section, large superclusters of galaxies lie at the ends of walls of galaxies. You can examine one of these superclusters in more detail with the following procedure.

22. Click **Home** and zoom outward, away from Sun's position in the Milky Way Galaxy to a distance of about **100 MLY.**

At this position, you can easily see the wall that you explored above, with a rich cluster of galaxies to the upper right of the view. This is the Virgo Cluster, as can be seen by clicking on the Highlight tab to reveal the Highlight pane, if this pane is not already open, and clicking on Virgo Cluster under the Supergalactic Eq. Plane layer.

It is possible to measure the distance between this cluster and the Sun by using the Location Scroller to align the cluster approximately with the Sun's position in the center of the screen.

23. Use the Location Scroller to move the Virgo Cluster to the center of the view window where its position coincides approximately with that of the Sun.
24. Left-click on any member of the Virgo cluster to open the Heads-Up Display (HUD), in which the distance to the chosen object is shown. The difference between the distances to the Sun displayed at the bottom of the screen and that to the member of the Virgo Cluster displayed in the HUD will be the distance between these two objects because they are both approximately on a single line of sight.

Question 1: What is the distance from the Sun to the Virgo cluster?

One of the large galaxies in the Virgo Cluster is Virgo A.

25. Select **Edit/Find** in the menu and enter **Virgo A** in the **Name Contains** box.
26. Use the Location Scroller to examine this galaxy from various angles in order to answer the following question.

Question 2: What is the Hubble classification of Virgo A?

This galaxy is very bright at radio wavelengths and contains a prominent jet emanating from a bright starlike core. It is obvious that some very energetic processes are going on in this core to produce this energy and the jet.

C. The Great Attractor

The galaxies in a large volume in the vicinity of the Milky Way appear to be moving towards a region of the sky filled with a relatively high density of galaxies that make up a complex structure. Because of this coordinated motion, this region has been called the **Great Attractor**. We can travel out to a distance from which we can view this as an entity, examine its structure and position and measure its distance from the Milky Way.

27. Click **Home** and zoom out to **300 MLY**.
28. In the Highlights pane, click on the **Great Attractor** heading and click off the **Show filament/group labels**.

You can see a huge pinwheel structure almost filling the view with a giant supercluster at its center. Many thousands of galaxies make up this structure, and the attractive force that this structure exerts upon the surrounding galaxies shows that there is a high concentration of mass there. You can examine this structure and measure its distance from the Sun using the Distance Indicator (DI) at the bottom of the view window and the HUD.

29. Use the Location Scroller to move the Great Attractor around to see its three-dimensional structure and locate its approximate center.
30. Click on any galaxy within the core of the Great Attractor to activate the Selective Heads-Up Display (S-HUD) that will show the distance of the chosen galaxy from your position.
31. Move the sky around to place the chosen galaxy in line with the Sun's position in the center of the view. One technique that can be used to accomplish this is to minimize the distance between your position and the chosen galaxy, displayed in the S-HUD. When this is at its minimum, the chosen galaxy is in line with the Sun in your view.

32. Determine the distance between the Great Attractor (GA) and the Sun by subtracting the distance to the GA, displayed in the S-HUD, from the distance to the Sun, displayed in the DI at the bottom of the view window.

Question 3: How far away from the Sun is the center of the Great Attractor?

Question 4: How does this result compare with the answer to Question 1? What conclusion can you draw from this result?

You can confirm your conclusion in Question 4 by using the Highlight facility in the highlight pane, displayed when you click on the Highlight tab.

33. In the Highlights pane, click Off the Highlight for the Great Attractor and click On the Highlight for the GA Virgo Cluster in the list under the Supergalactic Eq. Plane heading. You will see that the central feature of the Great Attractor is the Virgo Cluster, examined in a previous section, but that this concentration of mass is surrounded by a huge and complex structure of galactic walls.

In rotating the Great Attractor around in the sky, you will note that it has a long, linear feature. You can measure the linear extent of this overall structure.

34. Click the **Home** button and then zoom out to **250 MLY** from the Sun.
35. Highlight the **Great Attractor** again and click **Off** the **Show filament/group labels** to reduce the confusion in the view window.
36. Rotate the Great Attractor to place its long axis across the view.
37. Open the contextual menu on any galaxy at one end of the linear structure and select the Center option to place this outlying galaxy at the center of the view window. The DI at the bottom of the view window will now display your distance from this chosen galaxy below the distance to the Sun.
38. You can now rotate the Great Attractor to place the other end of the linear structure in front of the chosen galaxy so that galaxies at the extreme end of the linear structure are in an approximate straight line from your viewpoint. The nearby galaxies will move relatively fast across your field of view as you rotate the sky.
39. Click on any one of these nearby galaxies to activate the S-HUD. This will display your distance from this galaxy. Move the sky around until this distance is at a minimum, when this close galaxy will be on the line between the far galaxy and your position. The length of this structure is simply the difference between the distance to the centered galaxy, displayed in the DI at the bottom of the view window, and the distance shown in the S-HUD next to the chosen nearby galaxy.

Question 5: What is the approximate length of the linear structure of the Great Attractor?

Note that this linear structure contains the Virgo supercluster and that there is an almost continuous line of galaxies from end to end along this linear structure.

D. The Great Wall and other Colossal Walls

The final step in this Grand Tour of the extensive region of the universe surrounding our position takes you to the limit of the present Tully Collection of galaxies, to a distance where you can look in on these galaxies and discover the beautiful bubble-like shapes within which these galaxies reside.

40. Go **Home** and then zoom out to **400 MLY** from the Sun.
41. Highlight the **Great Attractor** and click **Off** the **Show filament/group labels**.
42. Rotate the sky around the Sun's position to examine one more time the limitation placed upon our overall view of the galactic distribution by the obscuration of large parts of the sky by our own galaxy, by finding the Zone of Avoidance across this galaxy collection in which no galaxies are seen.
43. Click **off** the highlight on the **Great Attractor** and highlight the **Great Wall** by clicking on the Great Wall heading in the highlight pane.
44. Zoom out to about **550 MLY.**
45. Rotate this view around the Sun's position using the Location Scroller to view this giant structure edge-on. Again, this wall is thin and relatively flat and extends over a huge distance, coincidentally appearing to be aligned in a direction that is perpendicular to the direction to the Sun.

This vast collection of galaxies is arranged on a surface surrounding a huge volume almost devoid of galaxies. Random processes cannot have formed this type of structure and these thin walls provide an important clue to the evolution of our universe following its origins in the Big Bang. There are other equally impressive structures stretching across this corner of the universe that have become apparent as astronomers have determined distances to galaxies and placed them in this three-dimensional mesh.

46. Highlight the **Southern Wall** in the Highlight pane to see another huge curtain of galaxies, somewhat closer to the Milky Way and smaller than the Great Wall and on the opposite side of the sky from our viewpoint. Rotate the sky to place this and the Great Wall edge-on to your view.
47. The distance between these giant walls can be measured by opening the contextual menu on a galaxy near the center of the Great Wall (e.g., in the Virgo Cluster) and choosing the Center command. The Distance Indicator at the bottom of the view window will now display your distance from this chosen galaxy.
48. Rotate the sky until the Great Wall is behind the Southern Wall and click on any nearby galaxy in this wall when it is near the center of the view. The S-HUD that appears will then display your distance to this galaxy. The spacing between these walls of galaxies is simply the difference between the distance to the centered galaxy shown on the DI at the bottom of the view window and that to the Southern Wall galaxy shown in the S-HUD.

Question 6:   What is the approximate spacing between the Great Wall and the Southern Wall?

You can explore this vast collection of galaxies at your leisure, highlighting features to guide you in this exploration and identification.

E. Beyond The Tully Collection

The Tully Collection of galaxies extends out to about 600 MLY and *Deep Space Explorer*™ provides you with the opportunity to examine the type and position of the full 28,000 galaxies in three-dimensional space. This remarkable data set is nevertheless limited and, whereas astronomers have a wide range of information about more distant objects, conclusions about these more distant realms are more speculative. *Deep Space Explorer*™ allows you to zoom outward into these regions and provides views and short essays to describe what we know about the spatial and evolutionary state of matter there.

49. Go **Home** and zoom out to about **700 MLY**, where you will have left the Tully Collection. This data set will be seen as a cube-shaped collection of galaxies but of course this is due to the limitations and extent of the collection. You can rotate this collection to find the Zone of Avoidance again and highlight the **Supergalactic Eq. Plane** to pinpoint our approximate position within this region of space and the galaxies with which we began this exploration. (Again, you can click **Off** the **Show filament/Group** labels to reduce the confusion in the view window.) The flatness of this wall of galaxies is particularly clear from this distant view.

50. Zoom outward to about **1950 MLY**. At this distance, *Deep Space Explorer*™ displays an image and a short essay on Galaxy Clusters.

Deep Space Explorer™ displays a Hubble Space Telescope image of a large cluster of galaxies known as Abell 2218. The significant thing about this image is the arclike structures surrounding the center, which are images of very distant galaxies imaged through a gravitational lens formed by matter within the foreground galaxy cluster. This type of lensing is one of the fascinating areas of current research in cosmology that is concerned with dark matter.

51. Zoom out to about **6000 MLY**. *Deep Space Explorer*™ displays a short essay and image of young galaxies, where the light has taken so long to arrive at Earth that these galaxies must be at an early stage in their evolution.

52. Zoom out to about **12,000 MLY** or 12 billion light-years, when *Deep Space Explorer*™ describes the Cosmic Microwave Background, the remnant radiation from the Big Bang that pervades the whole of space.

The measurement of this radiation and its structure is placing important constraints upon the theories of the evolution of the universe from the Big Bang onward. Interestingly, you can see evidence of this microwave radiation in your own home because photons from this source are detected by a television set when it is tuned to no specific station and thus many of the bright "noise" dots seen at this time are "signals" that have originated in the Big Bang!

53. Zoom out to the limit of the distance range in *Deep Space Explorer*™ to read the final short essay on the Big Bang that most cosmologists believe started the evolution of our present universe some 13 billion years ago.

54. As a final journey that helps to put into perspective the scale of distances that you have explored in this exercise, use the **Down** button to zoom back to the Solar System.

F. Conclusions

In this wide-ranging project, you have explored the large-scale structure of our universe in a way that is impossible in real life, using some of the best data in the world to travel vicariously though the soap-bubble features of the realm of the distant galaxies. You can investigate individual galaxies and collections and their characteristics and spacing at will by following the guidelines and methods outlined.

GLOSSARY OF PROCEDURES

This glossary outlines the software features most commonly used in the Observing Projects. References to the description of these features in the User's Guide are also given.

A. *Starry Night*™ Procedures

Angular Separation Measurement

The projects in this book make fairly extensive use of this feature and it is useful to become familiar with its use. To measure the angular separation between two objects, position the cursor over the first object and, when the cursor changes to an arrow shape, hold down the left mouse button and then drag the mouse to the second object. A line appears along with the angular distance between the first object and the current cursor position. You may find that this feature is quite sensitive to the initial position of the cursor and fails to engage on the first attempt so that, instead of measuring angular separation, movement of the mouse shifts the view of the sky. When this occurs, select **Edit/Undo Scroll** from the menu (or use the Ctrl + Z keyboard shortcut) to undo the change in view direction. Then try the angular separation measurement again. See page 21 of the User's Guide for more information on using this feature.

Centering an Object in the View

Open the contextual menu with the cursor over the object you wish to center in the view by right-clicking over it. Select **Center** from the menu. Using the Find facility to locate an object will center it automatically in the view and label it.

Changing Date and Time

You can enter specific dates and times in the control panel. Click on each field of the Date and the Time in turn to highlight it. Then enter the desired value for the field from the keyboard. You can also increment and decrement highlighted fields in the date and the time of the control panel with the + and – keys. See page 17 of the User's Guide.

Changing Viewing Direction

You can change the viewing direction with the **Hand Tool** by positioning the cursor on the background sky and holding down the mouse button while dragging the mouse. You can also select **View/Scrollbars** from the menu and use the scrollbars to change the viewing direction. The current viewing direction is described in the Status pane under the General/Looking item. See page 16 of the User's Guide for more details.

Changing Viewing Location

Change viewing location in the Viewing Location dialog window. There are several ways to access this dialog:

1) Select Options/Viewing Location... from the menu
2) Click on the current location in the control panel and select Other... from the menu that pops up
3) Use the keyboard shortcut Ctrl + L (Cmd + L on Macintosh)

> **TIP:** To scroll rapidly to a specific entry in the List pane, type the first letters of the desired location.

With the Viewing Location dialog window open, select the location from which you wish to view the sky from the List pane. Alternatively, you can specify the desired location by clicking on the map in the Map pane. Finally, you can specify a specific latitude and longitude as a location in the Latitude/Longitude pane. See Changing Your Viewing Location on page 67 of the User's Guide.

Contextual Menus

There are several types of contextual menus, depending upon the position of the cursor. When the cursor is over the background sky in the view and in the shape of a hand, clicking the right mouse button (Macintosh users hold down the Control key while clicking the mouse button) opens the Sky contextual menu. When the mouse cursor is over a specific object, this cursor changes to an arrow shape. Clicking the right mouse button opens the contexual manu. See page 48 of the User's Guide for more information on the Object Contextual Menu.

Finding Objects

Access the Find pane by selecting Edit/Find from the menu, opening the Find pane in the interface or using the keyboard shortcut Ctrl + F (Cmd + F on the Macintosh). Enter the name of the object you wish to find in the textbox at the top of the Find pane. When this textbox is empty, the Find pane shows the list of Solar System objects. See page 19 of the User's Guide.

Selecting Preconfigured Views

Display the preconfigured views requested in the projects by opening the **Go** menu and selecting the relevant view from the list under the appropriate **Chapter** in the list of **Observing Projects.**

Time Flow Controls

The Time Flow controls in the control panel allow you to move forward or backward in time smoothly or in discrete steps. The time step interval determines the rate at which time flows. In the projects, some of the instruction steps explicitly request single time steps forward or backward. See page 65 of the User's Guide for more information on the Time Flow controls.

Zooming in on Objects

Use the Field of View controls in the control panel to zoom in or out. You can zoom in to a field of view that is 6 arc minutes across using the slide bar. You must use the + or – keys to zoom to a smaller field of view. You can also use the wheel on a wheel mouse for this purpose if the cursor is over the main view. See page 20 of the User's Guide.

B. *Deep Space Explorer*™ Procedures
Contextual Menus

Click the right mouse button (Macintosh users hold down the Control key while clicking the mouse button).

Finding Objects

Select Edit/Find from the menu and type the name or designation of the object you wish to find in the edit box.

Go Home

Click the Home button to return to the default viewing location. This position is directly above and at a distance of 110,000 LY or 0.110 MLY from the position of the Sun on a line perpendicular to the Milky Way galactic plane.

Highlighting
You can highlight various features of the distribution of galaxies around the Milky Way by opening the Highlight pane and clicking on the selected structure.

Location Scroller
To activate the Location Scroller, click and drag the mouse in the view. This scrolling is most easily carried out by moving the mouse vertically or horizontally along the centerline of the screen to move the viewpoint around on a sphere centered upon the Sun the radius of which is displayed at the bottom of the view screen. You can change the inclination of the view by moving the mouse in small circles anywhere on the screen. This scrolling requires practice in order to move the view reliably and consistently in a specific direction. You can display Small or Large reference axes to help in this orientation by opening the Guides pane and activating Reference Axes.

Locking On/Centering Objects
To center an object in the view, position the cursor over the object, open the contextual menu, and select the Center command.

Selection HUD (S-HUD)
To activate the Selection HUD, lock on to the object as described in Locking On/Centering Objects, above.

Spaceship Mode
This mode allows you to explore the local universe around the Milky Way as if you are traveling in an intergalactic spaceship with capabilities well outside our present technologies and, indeed, well beyond physical possibility because you can simulate travel at impossible speeds, many orders of magnitude greater than the speed of light. The various controls and features of this mode are described in Chapter 7 of the User's Guide, starting on page 47.

Views
Opening the Views pane provides you with a wide range of specific objects and views of interest, ranging from local orbiting space stations, such as the Hubble Space Telescope and the International Space Station, through the planets, an overview of the positions of extra-solar planets, galaxies in galactic clusters, and on to views of the far reaches of our universe. The Location Scroller and Zoom controls allow you to move around each of these selected objects at a constant distance or move in closer to gain a three-dimensional perspective on these objects. A striking example of this facility is the view Inside the Virgo Cluster of galaxies, where several hundred galaxies are shown in their true three-dimensional positions.

Zooming
Use the down and up elevation buttons in the control panel to zoom in and out, respectively.